SPECIAL PAPERS IN PALAEONTOLOGY NO. 10

UPPER CRETACEOUS OSTRACODA FROM THE CARNARVON BASIN WESTERN AUSTRALIA

BY

R. H. BATE

With 27 plates, 42 text-figures,
and 2 tables

THE PALAEONTOLOGICAL ASSOCIATION
LONDON

JANUARY 1972

PRINTED IN GREAT BRITAIN

CONTENTS

ABSTRACT. Coniacian, Santonian, and Campanian ostracod faunas, comprising some 54 species, are described from the Carnarvon Basin, Western Australia. Of these, 9 genera and 32 species are new; 2 genera and 22 species, also new, are left under open nomenclature.

The ostracod fauna is identified as a Southern Hemisphere fauna, of which 11 genera are unknown in the north. A twelfth genus, *Oculocytheropteron* nov., is not yet confirmed as being entirely a southern form. 1 cosmopolitan genus, *Cytherella* Jones, is represented by 3 atypical species.

The stratigraphical value of the ostracods is shown and the lateral equivalence of the upper part of the Toolonga Calcilutite with the Korojon Calcarenite confirmed.

A new hinge type, characteristic of the Cytheruridae, is identified and named peratodont. 4 variations (holoperatodont, hemiperatodont, paraperatodont, and artioperatodont) are recognized.

INTRODUCTION

MARINE Cretaceous sediments crop out along the coastal region of Western Australia in the Carnarvon and Perth Basins (text-fig. 1), good stratigraphical accounts of which are to be found in Condon *et al.* (1956), McWhae *et al.* (1958), Belford (1959), and Condon (1968). In addition to the early work of Chapman (1917) on the Cretaceous foraminifera and ostracods of the Gingin Chalk, the work of Edgell (1954, 1957), Belford (1960), and others on the Cretaceous foraminifera of

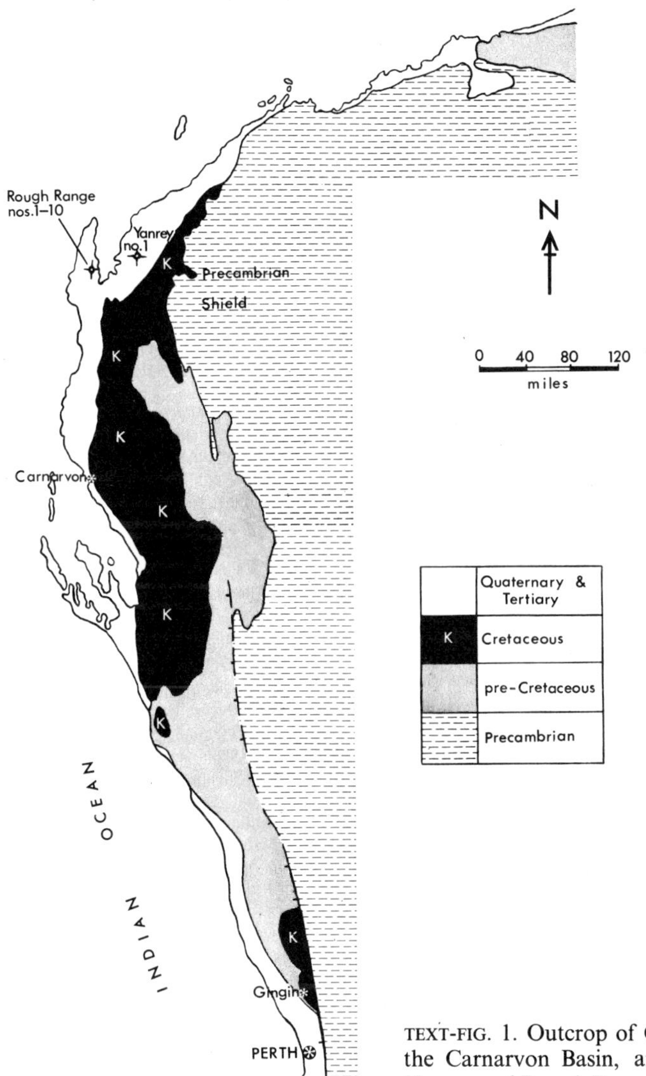

TEXT-FIG. 1. Outcrop of Cretaceous sediments in the Carnarvon Basin, and the location of the Yanrey and Rough Range boreholes. Map drawn from McWhae *et al.* 1958.

Western Australia has gone a long way towards accurately dating the lithological units concerned. The currently accepted dating of the Cretaceous sediments of the Carnarvon Basin, with some emendment to the upper limit of the Gearle Siltstone, is shown in Table 1.

The material for this investigation was provided by the West Australian Petroleum Pty. Ltd. from their Yanrey No. 1, Rough Range No. 4, and Rough Range South No. 1 boreholes. Altogether some 49 samples were examined, ranging from the Muderong Shale (Aptian) to the Korojon Calcarenite (Campanian). Of these only those listed below contained an ostracod fauna:

1. Rough Range South No. 1, core 53, 2330–2335 ft. Korojon Calcarenite, Campanian. Rich ostracod fauna.
2. Rough Range No. 4, core 1, 1380–1388 ft. Korojon Calcarenite, Campanian. Rich ostracod fauna.
3. Yanrey No. 1, core 3, 480–500 ft. Toolonga Calcilutite, Campanian. Rich ostracod fauna.
4. Rough Range South No. 1, core 59, 2435–2447 ft. Toolonga Calcilutite, Santonian. Rich ostracod fauna.

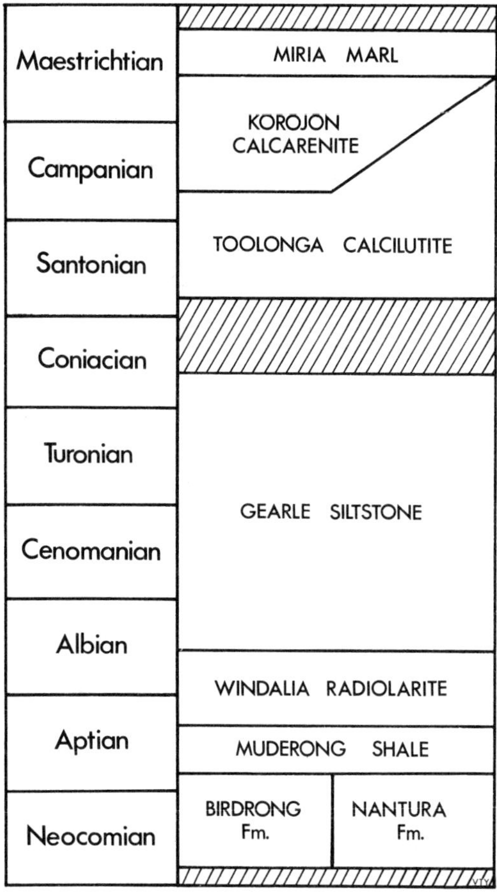

TABLE 1. Sequence of Cretaceous strata in the Carnarvon Basin.

5. Rough Range South No. 1, core 62, 2505–2511 ft. Upper Gearle Siltstone, Coniacian. Poor ostracod fauna.
6. Yanrey No. 1, core 4, 621–641 ft. Upper Gearle Siltstone, Coniacian. Poor ostracod fauna, fragments only.

The samples listed above have been numbered 1 to 6 and will be so referred to throughout the text.

The age dating of the samples is based on the planktonic foraminifera and was undertaken by Dr. D. D. Bayliss. The planktonic foraminifera as well as the ostracods (see Table 2) support the contention that the Korojon Calcarenite is merely a facies of the more widespread Toolonga Calcilutite. The first 3 samples are Campanian in age with no evidence of Maastrichtian faunas. Perhaps more interestingly, the samples of the Upper Gearle Siltstone possess a planktonic foraminiferan assemblage indicative of the Coniacian. This places the top of the Upper Gearle Siltstone much higher than suggested by previous workers (see Brown *et al.* 1968).

The purpose of the present paper is to describe the rich Campanian and Santonian ostracod faunas from the Carnarvon Basin. The Coniacian ostracods have also been included but these represent only a very minor part of the fauna as a whole.

During the investigation it was discovered that Dr. J. W. Neale was concerned with a revision of the F. Chapman ostracods from the Gingin Chalk, Perth. In order to undertake the study of the Carnarvon ostracods it was necessary to examine the Chapman type ostracods which are housed in the Bureau of Mineral Resources, Canberra. Because of Dr. Neale's interest in this material a revision of Chapman's ostracods will not be undertaken here; this will be conducted by Dr. Neale at a later date.

The ostracod terminology used throughout the text follows that of Moore (1961).

All illustrations have been produced by the author; photographic illustrations were taken on a Cambridge Instrument Stereoscan scanning electron microscope; text-figures were drawn using a Leitz camera lucida. Apart from Plate 1, fig. 6, all photographs are of specimens coated with gold under vacuum.

All measurements are inclusive of terminal spines.

Repository. The material on which this paper is based has been deposited in the collections of the British Museum (Natural History), specimens referred to being prefixed Io. Ostracods prefixed CPC. belong to the Chapman collection, housed in the Bureau of Mineral Resources, Canberra.

Acknowledgements. I am indebted to the West Australian Petroleum Pty. Ltd. for permission to publish on their material. I should also like to record my sincere thanks to Dr. Ross McWhae and Mr. M. H. Johnstone, who arranged for the material to be sent; to Mr. J. N. Casey (Bureau of Mineral Resources, Canberra), who kindly arranged the loan of Chapman's type ostracods from the Gingin Chalk; to Dr. D. D. Bayliss, who dated the Cretaceous samples; and to Miss Jill Houchen, who printed the photographs. I should also like to extend my sincere thanks to Dr. G. Deroo (Institut Français du Pétrole) for kindly sending me a number of specimens.

SYSTEMATIC DESCRIPTIONS

Order PODOCOPIDA Müller 1894
Suborder PLATYCOPINA Sars 1866
Family CYTHERELLIDAE Sars 1866
Genus CYTHERELLA Jones 1849

Remarks. 6 species of *Cytherella* are present in the Santonian and Campanian and 1 in the Coniacian of the Carnarvon Basin. Of these only 2 are sufficiently distinct and present in sufficient numbers for a specific name to be assigned. The remainder, represented by only a small number of individuals, are referred to simply as Types A–E respectively.

A large number of species of *Cytherella* has been described from Europe, North America, Asia, and elsewhere. The difficulty in accurately identifying species of this genus, based as they are on slight variations of carapace outline is reflected by the number of times authors tend to identify their species with those from another continent, often when other members of the ostracod fauna are not equally correlatable. The geographical range of most species does not appear to support an intercontinental correlation.

Cytherella atypica sp. nov.

Plate 3, figs. 1–4; text-figs. 2A, 2B, 2D, 3F

Diagnosis. Cytherella with ovoid outline; convex dorsal margin in female dimorph. Male dimorph with broad median angle about mid-point. Left valve larger than right. Female dimorph with 2 posterior pits on inside of valve.

Holotype. Io.4372, male left valve from the Toolonga Calcilutite, Sample 3 (Campanian).

Paratypes. Io.4370, 4371, 4373, 4374; 2 valves and 2 carapaces from Sample 3.

Other material. 22 valves and carapaces from Sample 3.

Description. Carapace oval, with broadly rounded anterior and posterior margins. Ventral margin convex with only a shallow median incurvature. Dorsal margin broadly convex, tending to be somewhat angular in outline at about mid-point. Female dimorph more oval and not so elongate as male, although in both, line of greatest length passes through mid-point. Greatest width and line of greatest height are both behind mid-point. Surface of carapace quite smooth.

Unusual feature of this species is the larger left valve, which overreaches right valve anteriorly but overlaps it elsewhere. Internally, left valve has broad contact groove extending around valve margin for reception of right valve selvage.

2 brood chambers or pits in posterior part of female dimorph, visible only from interior, and *not* reflected on outside by nodes.

Dimensions (in mm)

	Length	Height	Width
Holotype, Io.4372, male left valve	0·83	0·56	
Paratype, Io.4370, female right valve	0·79	0·50	
„ Io.4371, female left valve	0·83	0·58	
„ Io.4373, female carapace	0·85	0·59	0·40
„ Io.4374, male carapace	0·81	0·51	0·33

Remarks. The presence of 2 posterior pits or brood chambers in the female dimorph and the fact that the left valve is larger than the right make this species not only unlike any previously described *Cytherella* but also atypical for the genus as a whole.

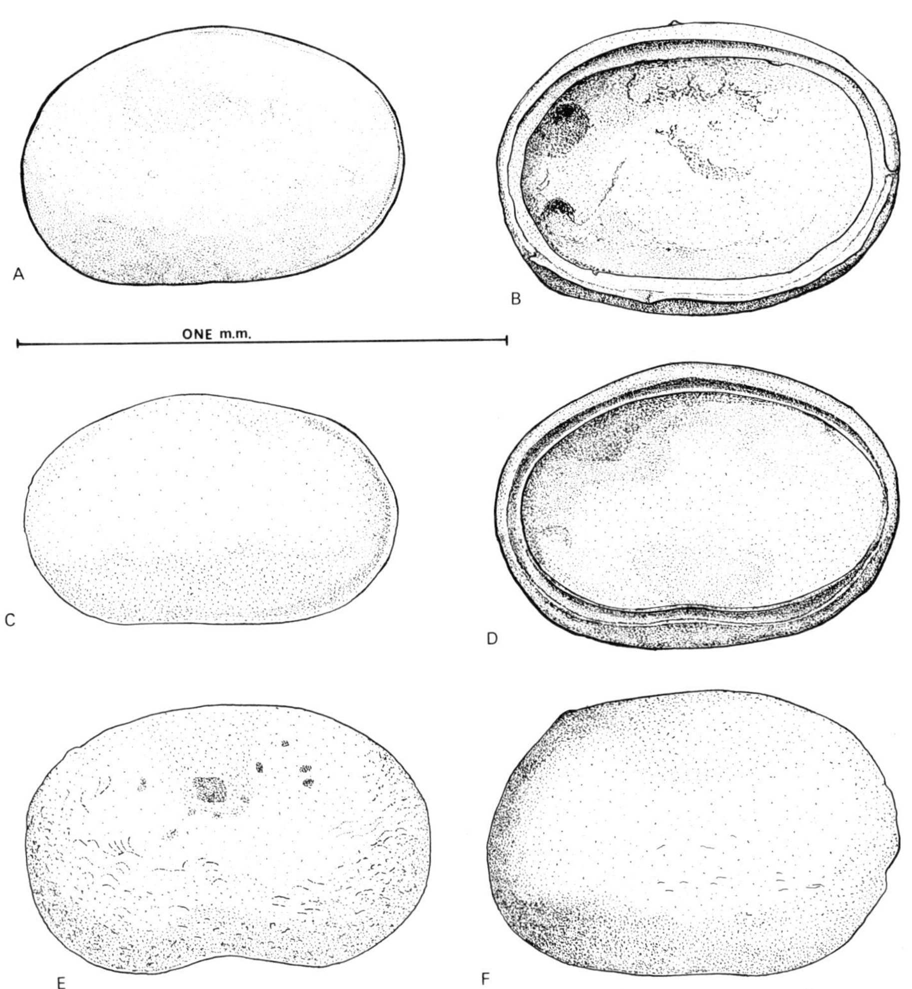

TEXT-FIGS. 2A–F. A, B, D, *Cytherella atypica* sp. nov.; A, female right valve, paratype Io.4370 (internal view in text-fig. 3F); B, internal view, female left valve showing 2 posterior brood pouches, paratype Io.4371; D, internal view, male left valve, holotype Io.4372. C, *Cytherella* sp. Type C, right valve, Io.4383. E, *Cytherella* sp. Type D, right valve, Io.4384. F, *Cytherella* sp. Type E, right valve, Io.4385.

Cytherella alata sp. nov.

Plate 3, figs. 6–8; text-figs. 3G, H

Diagnosis. Cytherella with oval/rectangular carapace in side view; spear-head shaped in ventral view. Postero-lateral part of carapace enlarged, alate; dorso-median depression distinct.

Holotype. Io.4375, carapace, Korojon Calcarenite, Sample 2 (Campanian).

Paratypes. Io.4376–4377, 1 left and 1 right valve, Sample 2.

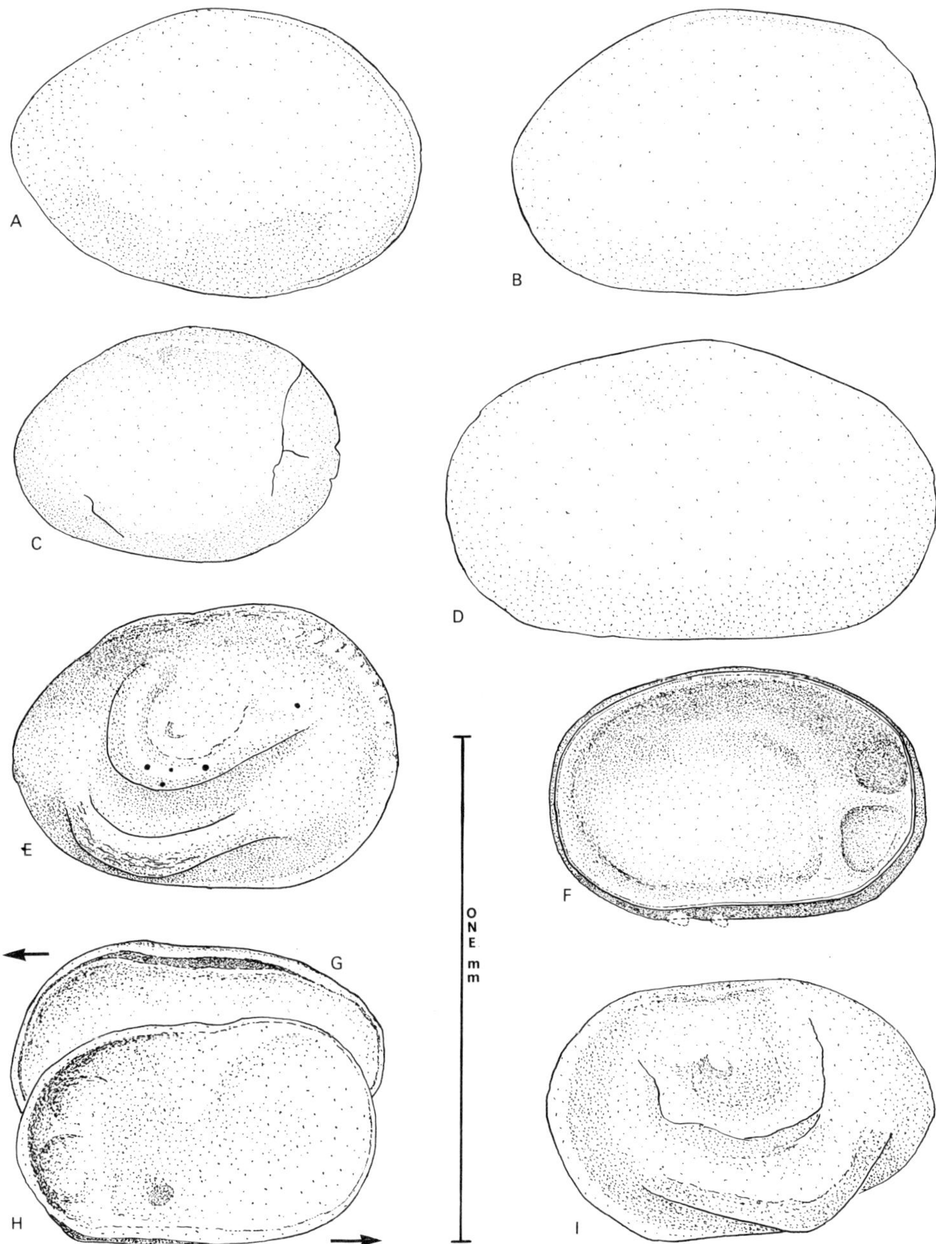

TEXT-FIGS. 3A–I. A, C, *Cytherella* sp. Type A; A, right valve, Io.4379; C, right valve, Io.4378. B, *Cytherella ovata* (of Chapman 1917), right valve, CPC.7149. D, *Cytherella muensteri* (of Chapman 1917), left valve, CPC.7148. E, I, *Cytherelloidea cobberi* sp. nov.; E, right valve, holotype Io.4386; I, left valve, paratype Io.4390. F, *Cytherella atypica* sp. nov., internal view, female right valve to show posterior brood pouches, paratype Io.4370. G, H, *Cytherella alata* sp. nov., internal views of right and left valves; note dorsal contact groove in right valve, paratype Io.4377, and posterior brood pouches in left valve, paratype Io.4376.

Description. Carapace rectangular/oval in lateral view with broadly rounded anterior end and more narrowly rounded posterior end. Dorsal margin gently concave above rather broad dorso-median depression. Postero-lateral part of valve develops an outward and ventrally projected alate outgrowth. Shell surface smooth, with rather large, circular, normal pore canal openings much in evidence.

Right valve larger than left, which it overlaps strongly along dorsal margin, less strongly along ventral margin. Terminally, virtually no overlap, and this is reflected internally by diminution of contact groove, which is only clearly developed in dorsal and ventral regions. Line of greatest length passes through or slightly below mid-point; greatest height lies in anterior third, and greatest width in posterior third.

Internally, valves rather shallow. In female dimorph, 2 weak brood chambers or pits at posterior end, but not reflected on outside of carapace. Dorso-median depression reflected internally as broad ridge, and depression marks position of alate outgrowths.

Dimensions (in mm)

	Length	*Height*	*Width*
Holotype, Io.4375, carapace	0·67	0·43	0·31
Paratype, Io.4376, female left valve	0·73	0·43	
„ Io.4377, male right valve	0·76	0·49	

Remarks. The alate development of *C. alata* is a unique character not repeated in any other species of this genus. Although the muscle scar pattern has not been observed there is no reason to doubt that *C. alata* is anything other than a true cytherellid.

Cytherella depressa Donze (1962) from the Oxfordian of France has some similarity of carapace outline but there is no real relationship with the present species.

Cytherella sp., Type A

Plate 3, fig. 10; text-figs. 3A, 3C

Remarks. 3 right valves of this species have been found in the Toolonga Calcilutite, Sample 4 (Santonian). So far no left valves have been found, making it impossible to name the species. On the outline of the right valve, however, there is some similarity to *Cytherella valanginiana* Neale (1962) from the Speeton Clay of England and to the line drawing of *Cytherella austinensis* Alexander as figured by Reyment (1960) from the Coniacian/Santonian of Nigeria. The species described by Chapman (1917) as *Cytherella ovata* Roemer (text-fig. 3B) from the Santonian Gingin Chalk has some similarity of outline, but tends to have a much more broadly convex postero-ventral slope and a larger antero-dorsal slope. The outline of these 2 Australian species is compared in text-figs. 3A–C.

Internally the brood pouches have not been clearly seen. There is a slight suggestion of 2 pouches being present, but this cannot be confirmed on the present material.

Dimensions (in mm)

	Length	*Height*
Io.4378, right valve	0·68	0·47
Io.4379, right valve	0·83	0·55

Cytherella sp., Type B

Plate 3, fig. 5

Remarks. 7 valves and 1 carapace have been obtained from the Toolonga Calcilutite, Sample 4 (Santonian), and 3 valves from the Korojon Calcarenite, Sample 1 (Campanian).

Compared with *Cytherella muensteri* Roemer of Chapman (1917), a Santonian species occurring in the Gingin Chalk, the present species is more rectangular in outline with a straighter dorsal margin in the smaller left valve. The female dimorph of this species, in common with the 2 new species described previously, has 2 internal posterior brood chambers, not reflected on the carapace exterior. This species is not, however, considered to be sufficiently distinct to justify the erection of a new specific name at this time.

Dimensions (in mm)

	Length	*Height*	*Width*
Io.4380, female left valve	0·69	0·40	
Io.4381, carapace	0·74	0·43	0·30
Io.4382, male right valve	0·62	0·37	

Cytherella sp., Type C

Text-fig. 2c

Remarks. 1 right valve present in the Upper Gearle Siltstone, Sample 5 (Coniacian).

Dimensions. Io.4383, right valve, length 0·76 mm, height 0·45 mm.

Cytherella sp., Type D

Text-fig. 2e

Remarks. 1 right valve, rectangular in outline with broadly rounded terminal margins, convex dorsal and ventral margins, has been found in the Korojon Calcarenite, Sample 1 (Campanian). This is a distinctive species which must remain unnamed due to the absence of additional material.

Dimensions. Io.4384, right valve, length 0·85 mm, height 0·51 mm.

Cytherella sp., Type E

Text-fig. 2f

Remarks. 1 rather massive right valve, of oval/rectangular outline, Korojon Calcarenite, Sample 1 (Campanian).

Dimensions. Io.4385, right valve, length 0·85 mm, height 0·57 mm.

Genus CYTHERELLOIDEA Alexander 1929
Cytherelloidea cobberi sp. nov.

Plate 1, figs. 7, 8; text-figs. 3E, 3I, 4A–C

Diagnosis. Cytherelloidea with oval carapace highest at anterior end. Central depression with 2 crescentic lateral ridges situated beneath, extended in postero-ventral direction, alate.

Holotype. Io.4386, right valve, Toolonga Calcilutite, Sample 3 (Campanian).

Paratypes. Io.4387–4396, 6 valves from Sample 3, 2 valves from the Korojon Calcarenite, Sample 2 (Campanian), and 1 valve from the Toolonga Calcilutite, Sample 4 (Santonian).

Other material. 13 valves from Samples 2 and 3, and 2 juvenile valves from Sample 4.

Description. Carapace oval, with maximum height at anterior end. Line of greatest length passes through mid-point in left valve, but passes below mid-point in right

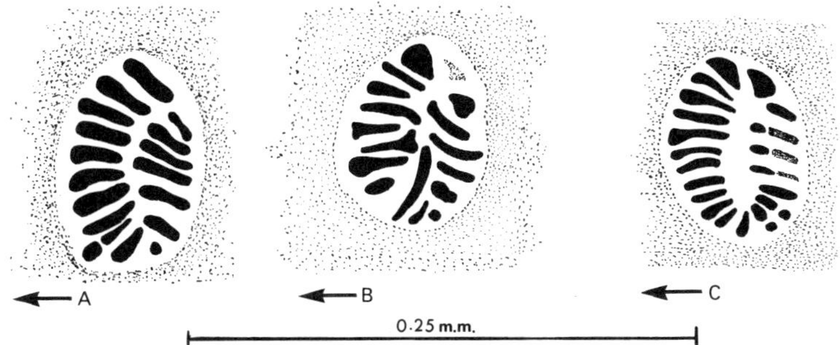

TEXT-FIGS. 4A–C. *Cytherelloidea cobberi* sp. nov. Variation in muscle scar patterns in left valves. A, paratype Io.4393; B, paratype Io.4392; C, paratype Io.4396.

valve. Although only disarticulated valves have been found, contact groove around periphery of right valve indicates that this valve is the larger. Dorsal margin gently concave medially, particularly in left valve. Anterior margin broadly rounded; posterior margin tapering, narrowly rounded. Ventral margin convex, with very shallow median incurvature.

Shell surface with broad and rather shallow dorso-median depression, with muscle scar configuration at centre. 2 crescentic ridges below depression with uppermost ridge bounding depression on 3 sides. Ridges project from valve in postero-ventral direction, producing an alate structure. Shell surface quite smooth, apart from some very fine wrinkles on lowermost ridge and below dorsal margin, and some small nodes around anterior margin.

Few, rather large normal pore canals on lateral surfaces of valves.

No signs of sexual dimorphism observed.

Muscle scars somewhat variable in arrangement of individual scars, consisting of biserial arrangement of 16 (possibly 17) to 20 elongate scars (text-figs. 4A–C), of which largest number is situated in the anterior facing row.

		Length	*Height*	*Width*
Paratype,	Io.4406, juvenile left valve	0·57	0·33	
,,	Io.4407, male right valve	0·91	0·56	
,,	Io.4408, juvenile carapace	0·52	0·32	0·21
,,	Io.4409, right valve	0·74	0·48	
,,	Io.4436, juvenile left valve	0·57	0·32	

Remarks. C. westaustraliensis ranges through sediments of Santonian and Campanian age in Western Australia. The specimens from the Gingin Chalk identified by Chapman (1917) as *Cytherella williamsoniana* Jones, *Cytherella williamsoniana* var. *stricta* Jones and Hinde, and *Cytherella chapmani* Jones and Hinde, are all female dimorphs of the present species and are not referable to the species to which Chapman assigned them. Indeed, the arrangement of the ridge ornamentation of *C. westaustraliensis* is completely different from that in the species mentioned.

In a number of individuals the ridge arrangement as indicated in text-figs. 5B, 5C, differs to the extent that the specimens concerned (text-figs. 5D, 5E) approach *C. cobberi* in appearance. The latter is unquestionably a distinct species, however, and is not a juvenile stage of *C. westaustraliensis*. Juveniles of *C. westaustraliensis* possess the adult female ornamentation. The presence of adult specimens having a similar ornamentation to *C. cobberi* suggests that the 2 species are closely related, possibly arising from a common ancestral stock.

The ornamentation as illustrated in text-fig. 5C is often modified in that the dorsal extension of the peripheral ridge dies out below the anterior cardinal angle and does not extend along the dorsal margin.

Ramsay (1968) suggested that a large number of Cretaceous and Tertiary cytherelloideans possess an ornament based on a spiral. In this context his *C. mairae* comes close to *C. westaustraliensis*, but differs in having a much broader ridge which spirals around the valve without any of the breaks found here. The ornamentation pattern found within the genus is an interesting one; the spiral arrangement could be interpreted as a breakdown of the *C. paraweberi* (Oertli 1957) type, where the peripheral ridge is completely unbroken and a short lateral ridge occurs at the valve centre.

Also of interest within the genus is the external development of the posterior nodes in the female dimorph. In many species, and *C. westaustraliensis* is a case in point, the posterior brood chambers, although well developed internally, are not reflected as nodes on the surface. The development of nodes is generally found in those species in which the ornamentation consists primarily of 2 or more lateral ridges extending from the posterior end. Other species, e.g. *C. cobberi*, apparently neither possess internal pits nor external nodes. The possibility here is that reproduction for these species is by parthenogenesis.

The problems to be worked on, as further species of *Cytherelloidea* are identified, concern the relationship of the 2 basic ornament patterns (spiral or circular and longitudinal lateral), and the interpretation of dimorphic characters.

Cytherelloidea carnarvonensis sp. nov.

Plate 2, figs. 4–9

Diagnosis. Cytherelloidea with rectangular carapace, broadly rounded at anterior and posterior ends, with deep marginal groove extending completely around free margin in male dimorph, interrupted posteriorly in female dimorph by 2 posterior swellings. Short lateral ridge projects forwards from each node; uppermost ridge extends to central muscle scar depression, and lowermost ridge into anterior half. In male dimorph central portion of lower ridge is below the muscle scar depression. Central area of valves strongly ornamented by large oval pits which are subdivided into smaller pits. Margin of valves, outside peripheral groove, ornamented by distinct wrinkles which run parallel with free margin. Distinct flange around anterior and antero-ventral margins.

Holotype. Io.4410, male carapace, Korojon Calcarenite, Sample 2 (Campanian).

Paratypes. Io.4411–4417, male and female valves and carapaces from the Korojon Calcarenite, Samples 1 and 2 (Campanian).

Description. Carapace rectangular, with broadly rounded terminal margins, concave ventral margin, and convex (antero-medially concave in female left valve) dorsal margin. Line of greatest length passes through mid-point, line of greatest height lies within anterior half, greatest width at posterior end.

Shell surface ornamented by rather large, oval pits subdivided by smaller pits. Central part of valve, apart from dorso-median depression, convex and separated from valve margin by prominent peripheral furrow, which in male dimorph extends completely around valve. In female dimorph furrow interrupted posteriorly by posterior swelling and development of 2 nodes which reflect the 2 internal pits. From each node, short lateral ridge extends forwards, uppermost ridge as far as central depression, and lowermost ridge into anterior half of valve. In male, only lowermost ridge present and does not extend to posterior end, being situated centrally below muscle scar depression. Surface of valves between peripheral furrow and free margin ornamented by distinct wrinkles which run parallel with valve edge.

Right valve larger than left and overlaps it on all sides. Shallow contact groove around inside of right valve for reception of selvage of left. Selvage in right valve as well developed as that in left, and to exterior of this, right valve has rather well-developed flange, seen in present material to extend around anterior and antero-ventral margins.

Muscle scars arranged in double row of approximately 12 elongate scars; precise number difficult to make out.

Dimensions (in mm)

	Length	*Height*	*Width*
Holotype, Io.4410, male carapace	0·66	0·40	0·24
Paratype, Io.4411, female carapace	0·68	0·42	0·28
,, Io.4412, male right valve	0·68	0·40	
,, Io.4413, female left valve	0·68	0·40	
,, Io.4414, female carapace	0·72	0·44	0·31
,, Io.4415, female right valve	0·73	0·44	

Remarks. C. carnarvonensis is unlike any described species of the genus. Only *C. sarsi* Puri (1960), a Recent species from the west coast of Florida, bears some similarity (as illustrated by Puri) in the nature of the surface pitting. Puri, however, described this ornamentation as a reticulation rather than pitting and there are 3 lateral ridges rather than 2 as here.

Genus PLATELLA Coryell and Fields 1937

Remarks. The genus *Platella* was erected by Coryell and Fields (1937) for cytherellids possessing an ornamentation of strong pits covering the entire shell surface. *Platella gatunensis* from the Miocene of Panama was selected as the type species. Subsequently *Platella jurassica* Bate (1963) from the Bajocian of England was placed in this genus.

Loetterle (1937) erected the genus *Morrowina* for species previously placed in *Cytherella* which had an ornamentation of weak punctae and thickened anterior and posterior marginal rims. *Cytherella excavata* Alexander from the Eocene of Texas was chosen as the type species. Crane (1965) included a strongly pitted form (unnamed) in this genus, but this is here regarded as belonging to *Platella.*

Reyment (in Moore 1961) and Van Morkhoven (1963) both placed *Morrowina* in synonymy with *Cytherella*. I agree with this, but cannot agree with Van Morkhoven in placing *Platella* also in synonymy with *Cytherella*.

The genus *Cytherella* has a large number of species differentiated on variations in carapace outline. The majority of the species are smooth shelled, but others, e.g. *Cytherella terminopunctata* Holden (1964), exhibit some degree of surface punctation. These are the *Morrowina* types and are more properly identified as *Cytherella*. No species of *Cytherella* are strongly ornamented over their entire surface; those having a strongly pitted shell surface belong to *Platella* and those with surface ribbing and swellings to *Cytherelloidea.*

Platella sp.

Plate 2, figs. 1, 2

Remarks. Only 2 left valves are available, from the Toolonga Calcilutite, Sample 3 (Campanian). The absence of additional material, especially of the right valve, precludes any specific assignment.

The shell surface is so strongly pitted that the ornamentation is more accurately described as reticulate. The small size of the valves suggests that these may be juvenile instars.

Dimensions (in mm)

	Length	*Height*
Io.4418, left valve	0·45	0·26
Io.4419, left valve	0·41	0·21

Suborder PODOCOPINA Sars 1866
Superfamily BAIRDIACEA Sars 1888
Family BAIRDIIDAE Sars 1888
Genus BAIRDIA McCoy 1844

Remarks. Sohn (1960) revised the Palaeozoic bairdiids and erected a number of new genera for those species which he considered did not belong to *Bairdia* s.s. Maddocks (1969) revised some of the Recent bairdiids and established a number of new genera for forms which diverge from the type. She did not comment, however, on the Recent species of *Bairdia* which morphologically might still be considered to be *Bairdia* s.s., a genus she clearly considered to be restricted to the Upper Palaeozoic (Maddocks 1969, p. 1). Sohn (1960, p. 13) rightly suggested that additional genera should be introduced for many of the post-Palaeozoic species and listed the morphological characters by which they should be distinguished. Of these the only character which appears to separate the smooth post-Palaeozoic bairdiids from the Palaeozoic forms is the presence of denticulate margins in some of the post-Palaeozoic species. The other characters he indicated refer to such other bairdiid genera as *Bairdoppilata* Coryell, Sample and Jennings (1935), *Triebelina* Bold (1946), *Neonesidea* Maddocks (1969), and *Paranesidea* Maddocks (1969). These genera appear to be perfectly valid, but we are still left with the smooth bairdiids which have a domed or angular dorsal outline, generally broad anterior margin with or without a concave postero-dorsal slope, and a tapered, generally slightly upturned, posterior end. That some species exhibit denticulate margins does not appear to warrant a generic distinction and I am not convinced that the phyletic history of the genus would be improved or made more accurate by erecting a new genus for these post-Palaeozoic forms.

Sylvester-Bradley (1950) showed that there is no justification in splitting the group of smooth bairdiids on the muscle scar pattern alone. Maddocks (1969) indicated that her genus *Neonesidea* is characterized by possessing a group of 8 elongate scars arranged in 3 oblique rows, whereas *Bairdoppilata* possesses the more typical rosette of 8 scars. It is suggested here that the basic grouping is a rosette of 8 scars, with some modification and variation occurring in individuals as well as between genera. In the species of *Bairdia* described here, 1 valve (Pl. 4, fig. 1) exhibits the typical rosette of both *Bairdia* and *Bairdoppilata*, whilst another valve has the *Neonesidea* arrangement (text-fig. 6). The presence of marginal denticles is also impersistent, appearing to be often restricted to large adults only.

TEXT-FIG. 6. *Bairdia austracretacea* sp. nov. Muscle scar pattern, right valve, paratype Io.4426.

The genera *Neonesidea*, *Triebelina*, and *Bairdoppilata* may be identified in fossil material and are quite distinct from the forms of which *Bairdia austracretacea* is a typical example. It is proposed to retain these forms in the genus *Bairdia* until such time as this arrangement is either confirmed or disproved by positive evidence.

Bairdia austracretacea sp. nov.

Plate 4, figs. 1, 2, 5, 8, 11, 12; Plate 5, fig. 6; text-figs. 6, 7A–I, 8

Diagnosis. Bairdia with high domed dorsal margin, concave antero-dorsal slope, convex postero-dorsal slope upturned at posterior end. Carapace strongly convex in dorsal view with greatest width median. Larger left valve with sinuous dorsal overlap of right valve. Shell surface smooth. Postero-ventral margin often dentate in adult instars.

Holotype. Io.4420, right valve, Toolonga Calcilutite, Sample 3 (Campanian).

Paratypes. Io.4421–4429, 5 valves and 1 carapace from Sample 3, and 2 valves and 1 carapace from the Korojon Calcarenite, Sample 1 (Campanian).

Other material. 4 valves from Sample 3, 6 valves and 2 carapaces from Sample 1, and 3 valves and 1 carapace from the Korojon Calcarenite, Sample 2 (Campanian).

Description. Carapace of large size with broadly arched dorsal margin and convex ventral margin; median concavity of left valve ventral margin rarely noticeable in side view. Anterior margin broadly rounded in left valve, antero-ventrally oblique in right. Posterior end acuminate, with upturned caudal process, and sometimes small denticles along convex postero-ventral margin. General outline of carapace variable, as shown in text-figs. 7A–I.

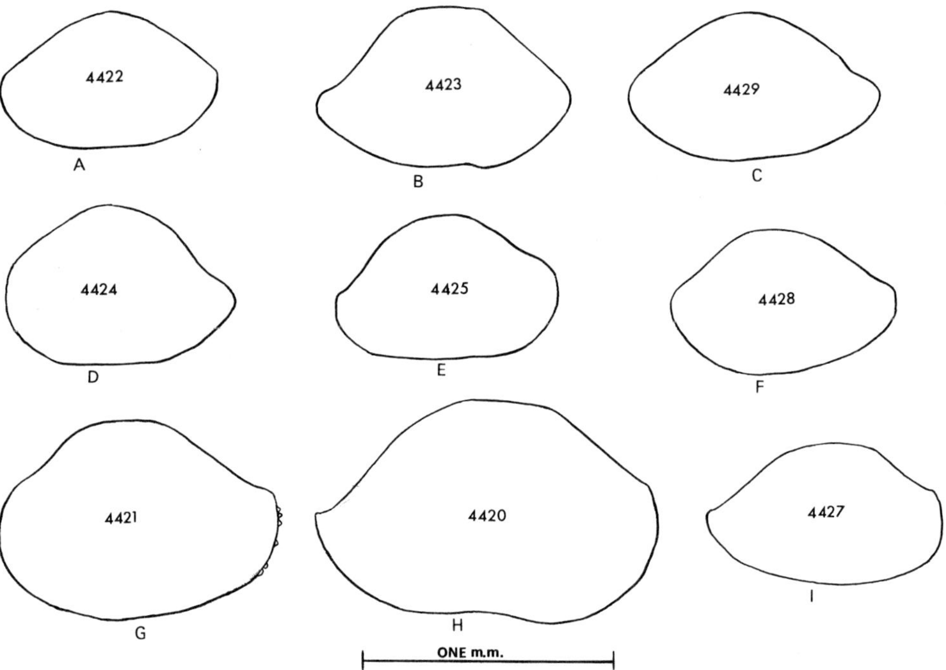

TEXT-FIGS. 7A–I. *Bairdia austracretacea* sp. nov. Variation in outline exhibited by individuals of the species.

Greatest length of carapace generally through mid-point in left valve, rarely below, but in right valve generally below. Greatest height and width both pass through mid-point. In dorsal view, overlap of right valve by larger left more strongly developed in antero-dorsal and postero-dorsal regions, producing a sinuous contact. Ventrally, left valve overlaps right most positively ventro-medially.

Shell surface smooth, normal pore canal openings small and widely scattered over surface. Hinge possesses 2 elongate terminal sockets in left valve, connected by very shallow depression having smooth bar beneath (text-fig. 8). Right valve hinge has antero-dorsal edges of valve developed to fit into terminal sockets of left valve, whilst low median ridge with shallow groove above is situated between. Inner margin

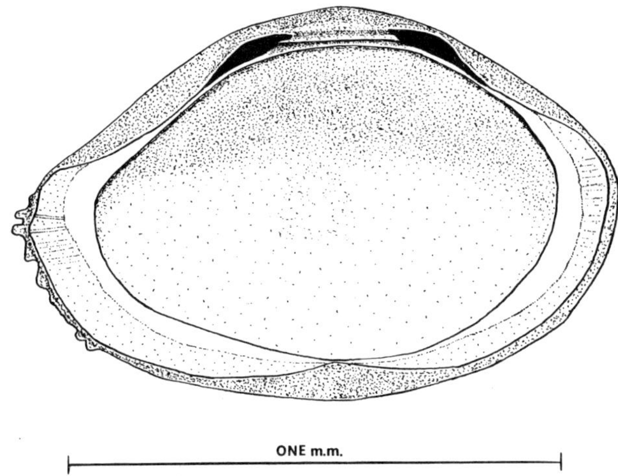

TEXT-FIG. 8. *Bairdia austracretacea* sp. nov. Internal view, left valve, to show hinge and structure of duplicature with anterior and posterior vestibules, paratype Io.4421.

and line of concrescence do not coincide and distinct vestibule developed, especially posteriorly and anteriorly. Radial pore canals short and straight; precise number not ascertained. Muscle scars appear either as circle of 7 elongate scars with an eighth scar at centre (Pl. 4, fig. 1) or as 8 elongate scars in cluster of 3 oblique rows (text-fig. 6). Both types possess 2 (?)mandibular scars antero-ventrally to main group.

Dimensions (in mm)

	Length	*Height*	*Width*
Holotype, Io.4420, right valve	1·37	0·83	
Paratype, Io.4421, left valve	1·17	0·75	
,, Io.4422, carapace	0·88	0·54	0·49
,, Io.4423, right valve	1·00	0·59	
,, Io.4424, left valve	0·93	0·62	
,, Io.4425, right valve	0·88	0·54	
,, Io.4427, right valve	0·90	0·51	
,, Io.4428, left valve	0·88	0·54	
,, Io.4429, carapace	0·95	0·56	0·45

Remarks. Van Veen (1934) described a number of Maastrichtian bairdiids from the Netherlands which have a close similarity in outline to the present species, e.g. *Bairdia cretacea, Bairdia decumana, Bairdia limburgensis,* and *Bairdia ubaghsi.* The variation in outline shown by *B. austracretacea* (see text-figs. 7A–I) brings it close to these species, especially to *B. decumana.* It differs, however, in that the line of greatest height passes through the mid-point rather than anterior to it, as in *B. decumana.* The antero- and postero-dorsal overlap of the right valve is also more exaggerated in *B. austracretacea,* a feature which clearly distinguishes it from Van Veen's other species, which have an almost straight line overlap in dorsal view. This feature also distinguishes this species from *B. subdeltoidea* var. *rotunda* Alexander (1927), which in dorsal view shows the strong median convexity of *B. austracretacea* but not the sinuous dorsal overlap. Furthermore *B. subdeltoidea* var. *rotunda* exhibits a punctate shell surface in some specimens, which is not a feature of *B. austracretacea.*

Genus BYTHOCYPRIS Brady 1880
?*Bythocypris* sp., Type A

Text-fig. 9A

Remarks. 3 complete carapaces, having an external outline close to that of *Bythocypris,* are present in the Toolonga Calcilutite, Sample 4 (Santonian). *Bairdia arquata* Chapman 1917 has some similarities of shell outline to this species, which differs by not being so tapered posteriorly or so strongly convex in dorsal view. The antero-dorsal and postero-dorsal slopes of ?*Bythocypris* sp. are also not so strongly angled as in *Bairdia arquata.*

Dimensions. Io.4445, carapace, length 0·78 mm, height 0·42 mm, width 0·29 mm.

?*Bythocypris* sp., Type B

Text-fig. 9B

Remarks. 4 complete carapaces from the Korojon Calcarenite, Sample 1 (Cam-

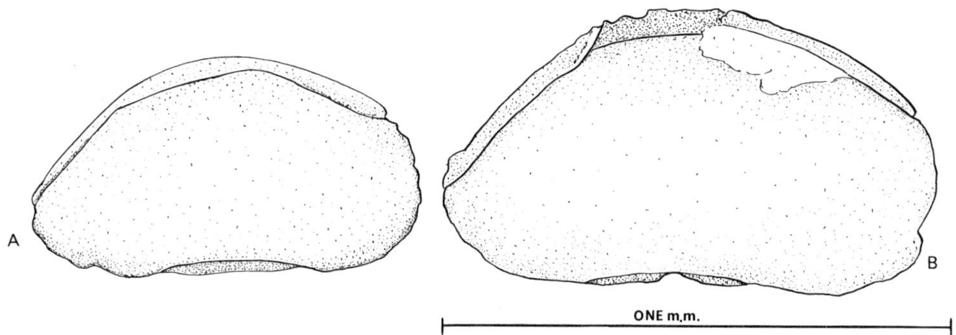

ONE m.m.

TEXT-FIGS. 9A, B. ?*Bythocypris* sp. A, right side of complete carapace, Type A, Io.4445; B, right side of complete carapace, Type B, Io.4446.

panian) represent a larger species than Type A, and although close to that species differ in size and by having a much shorter and more strongly convex dorsal margin.

Dimensions. Io.4446, carapace, length 0·98 mm, height 0·53 mm, width 0·32 mm.

Superfamily CYPRIDACEA Baird 1845
Family PARACYPRIDIDAE Sars 1923
Genus PARACYPRIS Sars 1866
Paracypris sp.

Text-fig. 10

Remarks. 2 right valves and 1 broken left valve from the Toolonga Calcilutite, Sample 3 (Campanian).

Chapman (1917) described a species of *Paracypris* from the Gingin Chalk which he identified as *Paracypris siliqua* Jones and Hinde. The species recorded here differs from Chapman's single figured specimen in that the dorsal margin is more strongly domed and the ventral margin more concave. The smaller right valve is not so strongly convex dorsally and approaches Chapman's species more closely, although the concavity of the ventral margin still serves to distinguish the 2. The number of specimens available, however, precludes a more definite identification of this material.

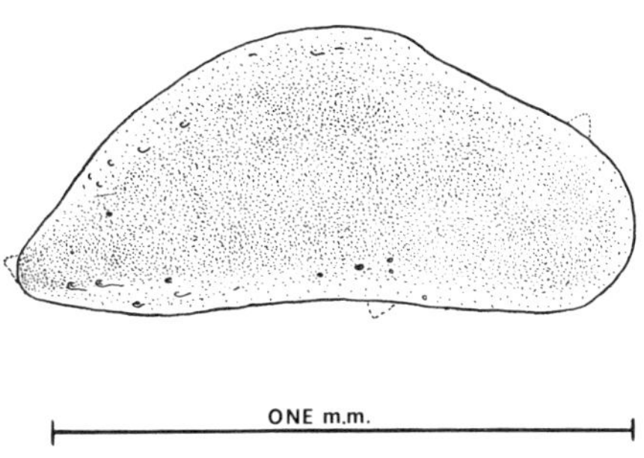

TEXT-FIG. 10. *Paracypris* sp., right valve, Io.4430.

Dimensions. Io.4430, right valve, length 1·13 mm, height 0·51 mm.

Genus PONTOCYPRELLA Ljubimova 1955

Remarks. The family assignment of *Pontocyprella* follows that of Moore (1961). Others, such as Neale (1962) and Mandelstam and Shneider (1963), placed the genus in the Pontocypridinae. It is most likely that both situations are incorrect. Purely for convenience it is proposed to follow Moore at present.

Pontocyprella dorsoconvexa sp. nov.

Plate 4, figs. 7, 10; text-figs. 11A–C

Diagnosis. Pontocyprella with elongate carapace; dorsal margin gently convex, becoming very slightly concave anteriorly in right valve. Ventral margin broadly concave. Greatest length of carapace below mid-point; posterior end situated close to ventral margin.

Holotype. Io.4440, right valve, Toolonga Calcilutite, Sample 3 (Campanian).

Paratypes. Io.4439 and Io.4441–4444, 4 valves from Sample 3 and 1 valve from the Korojon Calcarenite, Sample 2 (Campanian).

Other material. 5 specimens from Sample 2, and 2 specimens from Sample 3.

Description. Carapace elongate/rectangular, broadly rounded anteriorly, tapering posteriorly. Line of greatest length passes below mid-point with posterior extremity close to ventral margin. Dorsal margin broadly and gently convex, with slight anterior concavity in right valve which does not appear in left valve. Ventral margin broadly concave. Shell surface smooth.

Internally, hinge consists of simple elongate and rather narrow dorsal groove in left valve which articulates with slender, smooth dorsal selvage edge of right valve. Muscle scars consist of 3 rather large oval scars with a fourth situated close behind. These scars are situated in lower half of valve below mid-point. Inner margin and line of concrescence do not coincide anteriorly; very narrow vestibule present. It is possible, however, that much broader vestibule was present in this species at anterior and possibly also at posterior end. Appearance of duplicature suggests that delicate part forming vestibule has broken away. Straight, rather slender radial pore canals cross duplicature; precise number not determined.

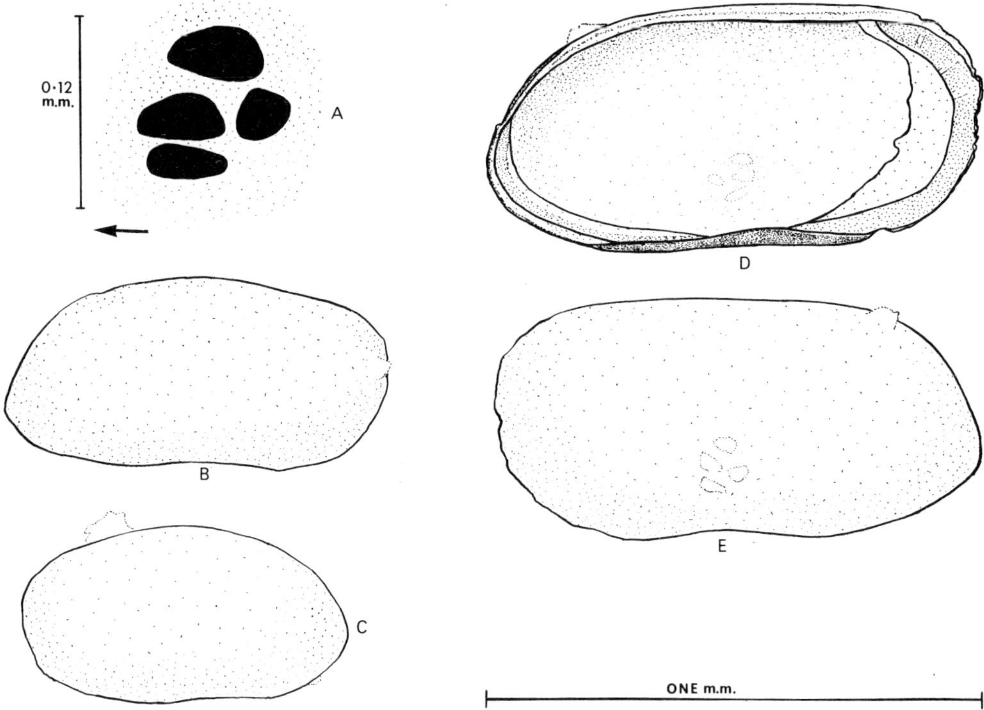

TEXT-FIGS. 11A–E. A–C, *Pontocyprella dorsoconvexa* sp. nov.; A, muscle scar pattern, left valve, paratype Io.4439; B, right valve, holotype Io.4440; C, juvenile left valve, paratype Io.4441. D, E, *Pontocyprella* sp., internal and external views, left valve, Io.4438.

Dimensions (in mm)

	Length	Height
Holotype, Io.4440, right valve	0·80	0·38
Paratype, Io.4439, left valve	0·83	0·41
„ Io.4441, juvenile left valve	0·66	0·36
„ Io.4443, right valve	0·78	0·39
„ Io.4444, right valve	0·85	0·40

Remarks. P. dorsoconvexa is close to *Pontocyprella harrisiana* (Jones), the type species, but is a more robust species, neither so slender in lateral outline nor so acuminate posteriorly. The convex dorsal margin of *P. dorsoconvexa* is especially well rounded in the region of the posterior cardinal angle, whereas in *P. harrisiana* it slopes more obliquely downwards and in the anterior part straightens out instead of being uniformly convex as in *P. dorsoconvexa. Pontocyprella maynci* Oertli 1958 has a more oval outline than *P. dorsoconvexa,* and *P. pertuisi* Donze 1964, again somewhat oval in outline, has a rather oblique slope to the dorsal margin.

    P. dorsoconvexa is associated in the Toolonga Calcilutite with a rare species of *Pontocyprella* which has almost parallel dorsal and ventral margins. Only 2 left valves have so far been found of this larger species, which exhibits a much broader anterior vestibule.

Pontocyprella sp.

Plate 4, fig. 6; text-figs. 11D, E

Remarks. 2 left valves have been found in the Toolonga Calcilutite, Sample 3 (Campanian). As the dimensions given below indicate, this species is much larger than the associated *P. dorsoconvexa.* The muscle scars for this species are the same as for *P. dorsoconvexa* and the type species, *P. harrisiana* (Jones).

Dimensions (in mm)

	Length	Height
Io.4437, left valve	0·93	0·45
Io.4438, left valve	0·94	0·45

Superfamily CYTHERACEA Baird 1850
Family BYTHOCYTHERIDAE Sars 1926
Genus BYTHOCERATINA Hornibrook 1953
Bythoceratina sp.

Text-fig. 12

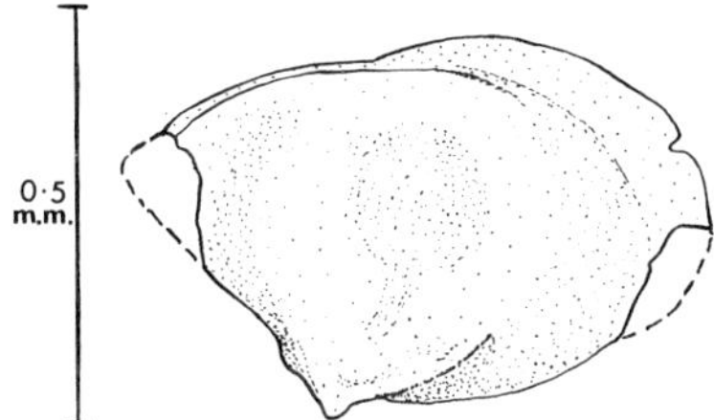

TEXT-FIG. 12. *Bythoceratina* sp., fragment of a right valve, Io.4431.

Remarks. 1 broken right valve present in the Korojon Calcarenite, Sample 2 (Campanian).

Dimensions. Io.4431, right valve, length (est.) 0·26 mm, height 0·43 mm.

Genus CYTHERALISON Hornibrook 1952

Remarks. Hornibrook erected this genus for 2 Recent ostracod species living off the coast of New Zealand, *Cytheralison fava* and *C. pravacauda*. In addition he also referred to the existence of fossil species (unnamed) occurring in the Eocene and younger sediments of New Zealand and in the Miocene of Australia. The record here of a Campanian species from Western Australia extends the geological range of this interesting genus, which has so far only been identified from Australasia.

Cytheralison has, until now, been retained in an incertae sedis family grouping. The examination of the muscle scars of specimens of *C. fava* deposited at the British Museum by Hornibrook confirms that the muscle scar grouping is an oblique row of 5 oval scars (text-fig. 13D) as found also in the genus *Bythoceratina*. Other characteristics of the carapace, such as the presence of a straight hinge line, a caudal process, and the internal division of the valve into a posterior and an anterior chamber by a median ridge (on which the muscle scars are situated), suggests that *Cytheralison* should be assigned to the Bythocytheridae.

Cytheralison contorta sp. nov.

Plate 11, figs. 12, 13; Plate 12, figs. 1, 2, 4, 8; text-figs. 13A–C

Diagnosis. Cytheralison with contorted lateral ridges extending from antero-dorsal region to postero-ventral region. Antero-dorsal and dorsal margins with frilled, upstanding ridge. Dorsal ridge in 2 parts; anterior end of posterior part turns down on to anterior part of lateral surface to fuse with lateral ridge. Development of dorso-lateral ridges lacking in some specimens.

Holotype. Io.4521, right valve, Toolonga Calcilutite, Sample 3 (Campanian).

Paratypes. Io.4522–4531. 9 specimens from Sample 3 and 1 specimen from the Korojon Calcarenite, Sample 2 (Campanian).

Other material. 8 specimens from Sample 3.

Description. Carapace quadrate, with dorsal and ventral margins almost parallel in adult instars; juvenile instars (Pl. 11, figs. 12, 13) have distinct posterior taper. Anterior margin of adult instars broadly rounded with strongly spinose flange, particularly well developed in antero-ventral region. Posterior end of carapace triangular or caudate with concave postero-dorsal slope and straight postero-ventral slope. Dorsal margin has frilled, upstanding ridge divided into anterior part which extends around antero-dorsal edge of anterior margin, and posterior half which dies out just behind posterior cardinal angle but which anteriorly passes down on to lateral surface of carapace. In some specimens this dorsal ridge extends for only short distance on to lateral surface, and in these cases development of lateral ridges is generally lacking. In specimens where dorsal ridge extends down on to lateral surface, ridge fuses with complex of contorted ridges which extend from antero-dorsal part of carapace to converge at strong postero-ventral projection common to all specimens. Ornamental variation is possibly dimorphic, with more highly ornate forms being males; this, however, remains to be proved. General shell surface has typical *Cytheralison* ornamentation of diamond-shaped cells covered by thin shell layer and opening to outside through narrow slit. Much larger slit marks position of muscle scars. Diamond pattern of cells obliterated by metallic coating necessary for examination under scanning electron microscope, but illustrated in text-fig. 13A by camera lucida drawing taken from uncoated specimen.

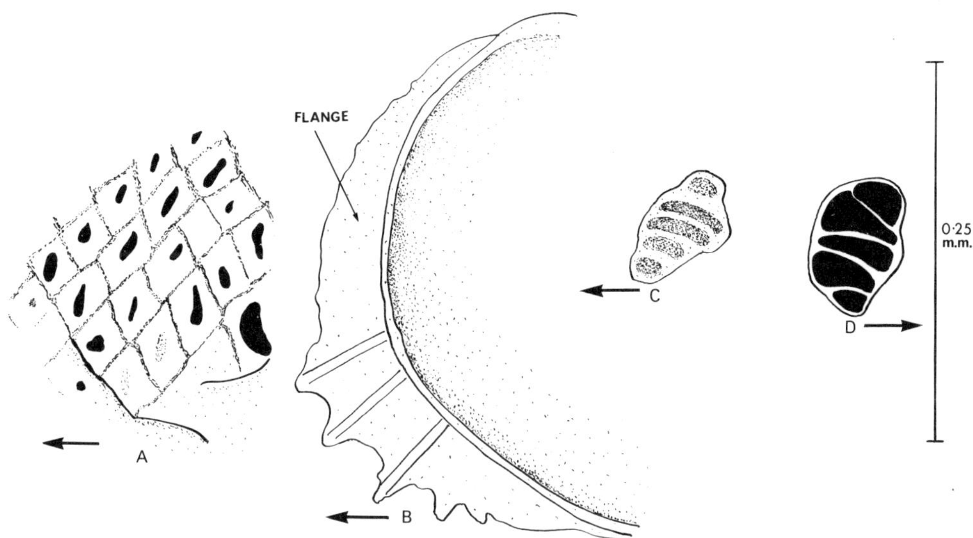

TEXT-FIGS. 13A–D. A–C, *Cytheralison contorta* sp. nov.; A, enlarged section of shell surface, showing cell covering with slit-like aperture, left valve, paratype Io.4527; B, anterior margin flange, right valve, paratype Io.4528; C, muscle scars, right valve, paratype Io.4528. D, *Cytheralison fava* Hornibrook (1952), muscle scars, right valve, Io.5008.

Internally, hinge of simple lophodont type, terminal posterior and anterior tooth of right valve being smooth terminations of selvage. Median groove long, narrow, smooth. In left valve, terminal sockets open to interior of valve, and hinge bar long, smooth. No accommodation groove, but dorsal margin set well back in left valve.

Selvage prominent in both valves, posteriorly cuts across caudal process, isolating it from central carapace chamber. Caudal process structurally continuous with flange, which is very broadly developed antero-ventrally, with spinose projections. Flange traceable as very low ridge above hinge, as broader structure along ventral margin, and therefore continuous around entire periphery of valve. Conspicuous groove at inside base of selvage (Pl. 12, fig. 8) extends around free margin of left valve, is continuous with the terminal sockets of left valve hinge, and is for reception of selvage of smaller right valve.

Inner margin and line of concrescence coincide to produce duplicature of moderate width. Radial pore canals not observed, although 3 (text-fig. 13B) seen to cross flange.

Muscle scars on vertical ridge which separates carapace chamber into 2 almost equal halves; difficult to determine. 1 specimen, however, (Io.4528, text-fig. 13C) has oblique row of 5 scars.

Dimensions (in mm)

	Length	*Height*
Holotype, Io.4521, right valve	0·85	0·45
Paratype, Io.4522, left valve	0·83	0·46
„ Io.4523, left valve	0·83	0·47
„ Io.4524, juvenile left valve	0·54	0·32
„ Io.4525, juvenile right valve	0·53	0·32
„ Io.4526, left valve	0·81	0·47
„ Io.4527, left valve	0·66	0·37
„ Io.4528, right valve	0·67	0·38
„ Io.4529, left valve	0·82	0·47
„ Io.4530, left valve	0·80	0·44
„ Io.4531, left valve	0·82	0·46

Remarks. The variation in surface ornamentation, which gives an impression, where the lateral ridges are lacking, of increased convexity, may be a dimorphic feature. Sexual dimorphism, however, has not so far been recorded in this genus.

C. *contorta* differs noticeably from the 2 Recent species (C. *fava* Hornibrook and C. *pravacauda* Hornibrook) in the possession of a strong postero-lateral projection and contorted lateral ridges. C. *fava* possesses a spine in approximately the same position as the postero-ventral lateral projection of C. *contorta*, but like C. *pravacauda* completely lacks lateral ridges.

Genus MONOCERATINA Roth 1928
Monoceratina invenusta sp. nov.

Plate 5, figs. 3–5; text-figs. 14A–D

Diagnosis. Monoceratina with warty appearance; anterior, dorsal, postero-ventral margins denticulate. Lateral surface with large nodes, 2 of which (1 anterior to and 1 posterior of a shallow median sulcus) may develop into lateral spines. Numerous small nodes and spines scattered over granular shell surface.

Holotype. Io.4432, right valve, Korojon Calcarenite, Sample 2 (Campanian).
Paratypes. Io.4433–4435, 2 right valves and 1 left valve from Sample 2.

Description. Carapace rectangular with long, straight dorsal margin. Anterior end broadly rounded, posterior end caudate, extreme end of caudal process being flattened. Greatest length of carapace passes just below dorsal margin, greatest height in anterior half of right valve and in posterior half in left valve. Postero-ventral margin in both valves broadly convex, with small denticles. Bordering denticles present along dorsal margin and around anterior margin. Shell surface granular, with very small spines and general warty appearance due to irregular development of nodes. Shallow median sulcus just visible, with large node complex situated to front and rear. Postero-ventral node tends to develop into prominent spine, as may antero-ventral one.

Hinge consists of long thin dorsal bar in left valve fitting into long narrow groove in right valve.

Dimensions (in mm)

	Length	Height
Holotype, Io.4432, right valve	0·71	0·32
Paratype, Io.4433, left valve	0·55	0·30
„ Io.4434, right valve	0·77	0·32
„ Io.4435, right valve (broken)	0·73	0·27

Remarks. M. invenusta is similar in appearance to *M. letopa* Gründel (1964) from the Upper Albian of East Germany. In dorsal outline these 2 species may, however, be easily distinguished; *M. letopa* has a much slimmer carapace, a more positive

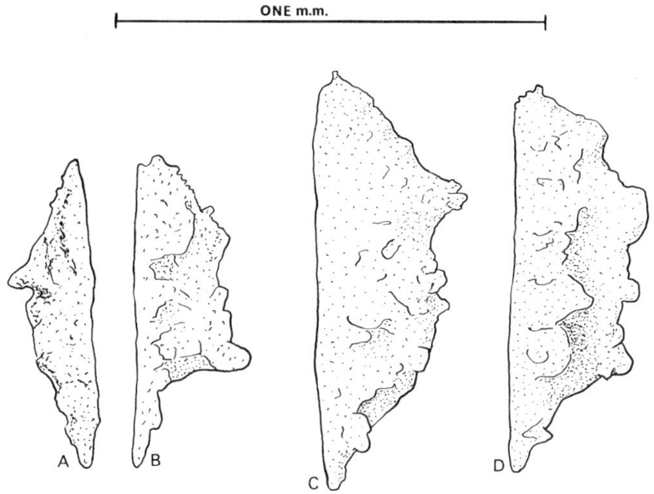

TEXT-FIGS. 14A–D. *Monoceratina invenusta* sp. nov., dorsal views of left and right valves to show variation in outline and ornamentation. A, left valve, paratype Io.4433; B, right valve, paratype Io.4435; C, right valve, paratype Io.4434; D, right valve, holotype Io.4432.

median sulcus and a thin marginal border which is not present in the more solid *M. venusta. M. bicuspidata* Gründel (1964) from the Middle Albian of East Germany has a closely similar lateral outline to *M. invenusta* but is not so robust in dorsal view and possesses rather better developed lateral spines.

Family PROGONOCYTHERIDAE Sylvester-Bradley 1948
Subfamily PROGONOCYTHERINAE Sylvester-Bradley 1948
Genus MAJUNGAELLA Grekoff 1963

Remarks. Majungaella was first described from the Upper Jurassic and Lower Cretaceous of the Malagasy Republic by Grekoff (1963) and subsequently (as *Neocythere*) from the Neocomian of South Africa (Dingle 1969*a*). I have also found species of this genus from the Callovian to Kimmeridgian of Tanzania and a new species from the Upper Albian–Lower Cenomanian of Tanzania has been described

(Bate 1969). Since it possesses a large number of stratigraphically short ranging species, *Majungaella* is proving to be a useful index fossil in Africa and may prove equally useful in Australia.

Majungaella is very close to *Progonocythere* Sylvester-Bradley (1948), but differs primarily in the possession of an increased number of anterior radial pore canals. *Majungaella annula* from Australia has a greater number of anterior radial pore canals than *M. perforata*, the type species from the Portlandian, Malagasy Republic. There is thus an increase in the number of anterior radial pore canals from the Upper Jurassic to the Cretaceous. This closely parallels the situation found in a number of lineages, e.g. *Eoschuleridea–Schuleridea–Aequacytheridea*. A new generic or sub-generic name for the Australian species of *Majungaella* is not considered necessary at this time.

Majungaella is possibly restricted to a different climatic zone than the closely related European genus *Progonocythere*. Records of *Progonocythere* outside Europe are considered to be doubtful, and so far *Majungaella* would appear to be restricted to a circum-Indian Ocean distribution.

Neocythere Mertens (1956) is a genus having a close external similarity to both *Progonocythere* and *Majungaella*. According to Kaye (1963*a*) an entomodont hinge as present in the last 2 genera is not present in the 3 subgenera of *Neocythere*, and both may further be distinguished by the outline of the posterior end, which is of blunted outline with a convex postero-dorsal slope in *Progonocythere* and *Majungaella*, and triangular with a concave postero-dorsal slope in *Neocythere* (text-figs. 15A, B). The position of the posterior hinge socket is also very much closer to the posterior end in the *Progonocythere/Majungaella* group.

Majungaella annula sp. nov.

Plate 5, figs. 7–9; Plate 6, figs. 2–7; Plate 7, figs. 1–4; text-figs. 15B, 16A, B

Diagnosis. Majungaella with pyriform carapace outline, postero-dorsally upturned. Shell surface with concentric pattern of low ridges, more prominently developed in ventro-lateral and posterior parts of carapace. Small marginal denticles often apparent.

Holotype. Io.4450. Carapace, Toolonga Calcilutite, Sample 4 (Santonian).

Paratypes. Io.4451–4464. 7 specimens from Sample 4, and 7 specimens from the Toolonga Calcilutite, Sample 3 (Campanian).

Other material. 6 specimens from Sample 4, 11 specimens from Sample 3, and 8 specimens from Sample 2 (Campanian).

Description. Carapace pyriform, with steeply backwardly sloping dorsal margin posteriorly upturned, degree of upturning varying with individuals. Anterior margin high, broadly rounded; posterior margin bluntly and narrowly rounded. Ventro-lateral margin strongly convex, overhanging ventral surface. Carapace convex in dorsal view, with broad, flattened, anterior marginal border, and narrower, flat-tened, posterior border. Small marginal denticles often apparent on posterior and anterior margins, especially in antero-ventral and postero-ventral regions. Denticles may be absent, depending on state of preservation. Left valve larger than right, with distinct ventral overlap.

Shell surface ornamented by low ridges arranged in concentric pattern, but only clearly seen in posterior half. Anteriorly, ornamentation fades away. Towards centre of valve, concentric ridges break down to form weak reticulate structure, whilst on ventral surface 3 prominent parallel ridges on each valve. Normal pore canals open to lateral surface with concentric alignment as for ridges.

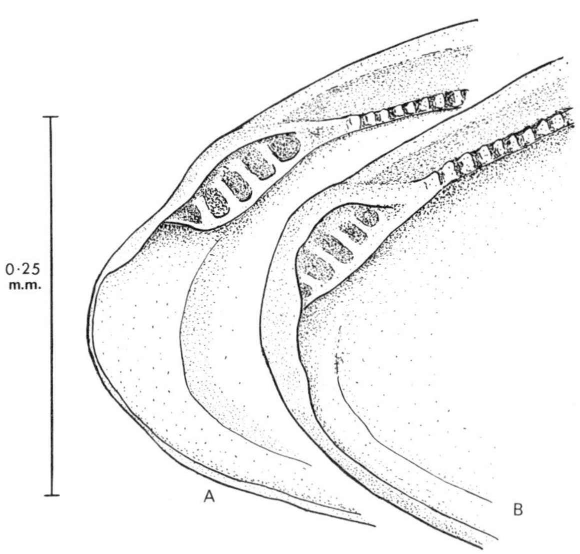

TEXT-FIGS. 15A, B. Posterior end, left valves, to show the contrasting outline and situation of the posterior hinge socket in *Neocythere* and *Majungaella*. A, *Neocythere vanveeni* Mertens, Gault, Folkestone; B, *Majungaella annula* sp. nov., Santonian, W. Australia.

Internally, valves articulate by means of strong entomodont hinge, having in right valve 4 large and 2 small anterior teeth and 7 posterior teeth; median groove coarsely loculate and expanded in its anterior part. Complementary structures present in left valve.

Inner margin and line of concrescence coincide to produce broad duplicature. Anterior radial pore canals (text-fig. 16B) long, slightly curved, sometimes branching, numbering 28–30. Only 5 posterior radial canals (text-fig. 16A). Some specimens exhibit rows of fine striae on duplicature, parallel to valve margin.

Muscle scars consist of 4 oval adductor scars arranged in crescentic row, with oval antero-dorsal frontal scar. Small scar to anterior of frontal scar (Pl. 6, fig. 5).

Dimensions (in mm)

	Length	Height	Width
Holotype, Io.4450, carapace	0·77	0·52	0·42
Paratype, Io.4451, ,,	0·69	0·48	0·39

C

		Length	Height	Width
Paratype, Io.4453, carapace		0·72	0·50	0·43
„	Io.4455, left valve	0·72	0·50	
„	Io.4456, left valve	0·73	0·54	
„	Io.4457, right valve	0·69	0·48	
„	Io.4459, left valve	0·71	0·53	
„	Io.4460, right valve	0·77	0·51	
„	Io.4461, left valve	0·77	0·54	
„	Io.4462, right valve	0·77	0·51	
„	Io.4463, right valve	0·72	0·51	
„	Io.4464, left valve	0·74	0·52	

Remarks. The ostracod from the Gingin Chalk incorrectly identified as *Cytheropteron concentricum* by Chapman (1917) belongs to *Majungaella,* but although closely

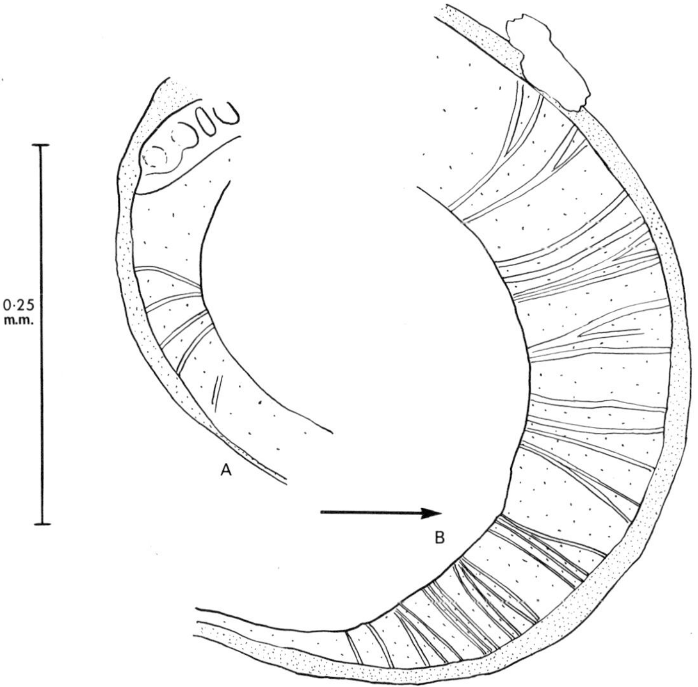

TEXT-FIGS. 16A, B. *Majungaella annula* sp. nov., posterior and anterior duplicatures showing radial pore canals, left valve, paratype Io.4456.

similar to *M. annula* differs in its distinct reticulate ornament, more convex dorsal margin, and less tapered posterior outline. The rather weak ornament of annular ridges clearly distinguish *M. annula* from the known species of this genus.

The considerable increase in the number of anterior radial pore canals when compared with the number present in the type species, *M. perforata* Grekoff (1963), is accepted as being within the possible range for the genus. A subgeneric rank may, however, become necessary for these forms as additional species are identified.

M. annula ranges from the top(?) of the Santonian into the Campanian. There does not appear to be any change in size throughout this range.

Family SCHIZOCYTHERIDAE Mandelstam 1960
Subfamily SCHIZOCYTHERINAE Mandelstam 1960

Remarks. Previously (Bate 1967) I had suggested that the Schizocytheridae were possibly a polyphyletic group based purely on the presence of a schizodont hinge. Hanai (1970) showed that the schizodont ostracods belong to the same subfamily group, despite the degree of variation in carapace outline and ornamentation. Hanai, however, divided the Schizocytherinae into 2 tribes, the Schizocytherini Mandelstam 1960 and the Paijenborchellini Deroo 1966.

Whilst agreeing with Hanai in his assessment of the schizodont ostracods I prefer to broaden the concept of the Schizocytherinae to include genera having an anti-merodont hinge, e.g. *Cnestocythere*, *Acrocythere*, and *Apateloschizocythere*. In doing this I consider it necessary to raise the Schizocytherinae to family status and the 2 tribes to subfamily level. Accordingly the subfamily Schizocytherinae of this paper is equivalent to the tribe Schizocytherini of Hanai (1970).

Hanai included only 2 schizodont ostracods in the Schizocytherini (herein Schizocytherinae), namely *Schizocythere* and *Amphicytherura*. A third schizodont ostracod (herein raised to generic level), *Sondagella* Dingle 1969*b*, can also be included. A number of ostracods which possess similar characters to the above with respect to carapace outline, radial pore canals, sieve plates, muscle scars, and shell ornamentation are also considered to belong to the subfamily even though their hinge dentition differs. The following ostracods are, therefore, considered to belong to the Schizocytherinae:

> *Schizocythere* Triebel 1950
> *Cnestocythere* Triebel 1950
> *Acrocythere* Neale 1960
> *Amphicytherura* Butler and Jones 1957
> *Apateloschizocythere* gen. nov.
> *Sondagella* Dingle 1969*a*.

The geographical range of the Schizocytherinae, restricted to the northern hemisphere by Hanai (1970, p. 715) is thus extended to include South Africa (*Sondagella*) and Australia (*Apateloschizocythere*). Similarly the geological range of the subfamily is extended into the Neocomian by the inclusion of the southern hemisphere *Sondagella* and the northern hemisphere *Acrocythere*.

Genus APATELOSCHIZOCYTHERE nov.

Type species. Apateloschizocythere geniculata sp. nov.

Diagnosis. Carapace quadrate with horizontal and oblique lateral ridges and frilled reticulate ornament. Hinge well developed, antimerodont. Eye node absent.

Duplicature broad; anterior radial pore canals long, straight, few in number. Normal pore canals with round to oval sieve plates, each bearing small nodes, numerous small pores, and single larger hair pore.

Remarks. Apateloschizocythere externally closely resembles *Schizocythere* but differs by lacking an eye node and in the possession of an antimerodont hinge. The otherwise close similarity of these 2 forms, however, suggests a phylogenetic relationship. It is interesting to note that the sieve plates present in *A. geniculata* (Pl. 15, fig. 7) are precisely the same as illustrated for *Schizocythere kishinouyei* (Kajiyama) by Hanai (1970, p. 701), even to the extent of possessing small nodes on the plate itself. *Cnestocythere*, also closely similar to *Apateloschizocythere*, differs in the possession of a distinct eye node and a dentate marginal projection above the posterior hinge socket of the left valve. *Acrocythere*, a more elongate genus, is close to *Apateloschizocythere* in all features other than carapace outline. *Amphicytherura* and *Sondagella* differ in the possession of an eye node and a schizodont hinge.

Apateloschizocythere geniculata sp. nov.

Plate 7, figs. 5–8; Plate 8, figs. 1–10; Plate 15, fig. 7; text-figs. 17A, B

Diagnosis. Apateloschizocythere with prominent oblique median ridge bent as in a knee joint.

Holotype. Io.4465, left valve, Toolonga Calcilutite, Sample 3 (Campanian).

Paratypes. Io.4466–4479+4679, 15 specimens, Sample 3; and Io.4480, 6 specimens, Toolonga Calcilutite, Sample 4 (Santonian).

Other material. 14 specimens from the Korojon Calcarenite, Sample 1 (Campanian), and 19 specimens from the Korojon Calcarenite Sample 2 (Campanian). 15 specimens from Sample 3 and 15 specimens from Sample 4.

Description. Carapace quadrate with broadly triangular posterior end; straight dorsal margin overreached by frilled dorsal ornamentation. Anterior margin rounded; ventral margin convex with median incurvature. Acutely angled cardinal angles enhance pronounced quadrate outline of carapace.

Apart from frilled reticulate ornament, 3 prominent lateral ridges arise from anterior margin. Lowermost ridge cuts down to extend back along ventral surface, whilst one above has ventro-lateral position, terminating in postero-ventral extension from carapace. Uppermost ridge starts below mid-height on anterior margin, extends back for $\frac{1}{3}$ length of valve where sharp bend occurs, and then follows oblique postero-dorsal route. Other features of carapace ornamentation seen in 4 left valves and 2 right valves illustrated; these demonstrate absence of variation within species.

Normal pore canal openings small, but each has sieve plate ornamented by small nodes (Pl. 15, fig. 7) and bearing small pore openings [Sp.] and single peripheral hair pore [Hp.].

Valves virtually equivalve, only noticeable overlap being of right valve by left in region of anterior cardinal angle. Dorsal flange of right valve overlaps dorsal margin of left. Internally, in left valve, groove inside selvage accepts selvage of right valve in anterior, posterior, and ventral regions. In this respect, left valve is larger than right. Flanges butt against each other without overlap.

Hinge strongly developed, antimerodont. Internally, valves divided by low median ridge, not represented externally by sulcus. Muscle scars not observed. Inner margin and line of concrescence coincide around free margin except for anterior region, where there is slight extension inwards of calcified inner lamella, but not sufficiently to produce vestibule. Duplicature broad; 10 long, straight anterior radial pore canals (at least 3 false, opening to lateral surface) and 4 posterior radial pore canals (text-figs. 17A, B).

Selvage in both valves well developed, that of right valve fitting into groove situated inside left valve selvage. Broad flange extends completely around each valve;

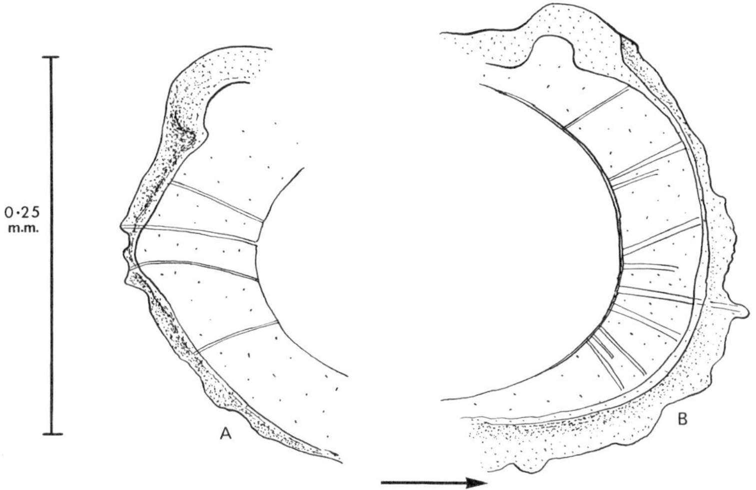

TEXT-FIGS. 17A, B. *Apateloschizocythere geniculata* gen. et sp. nov., posterior and anterior duplicatures showing radial pore canals, selvage and flange, left valve, holotype Io.4465.

in left valve, however, present dorsally as very low ridge; in right valve strong dorsal development of flange seen in Plate 8, figs. 2, 3, 10.

Eye node not developed, although shallow internal depression below anterior hinge teeth of right valve is of similar development and situation to internal eye node depression of *Acrocythere*. Here depression does not lead to eye node but instead has 2 normal pore canals leading from it. Depression not present in left valve.

Dimensions (in mm)

	Length	Height	Width
Holotype, Io.4465, left valve	0·49	0·30	
Paratype, Io.4466, right valve	0·51	0·32	
,, Io.4467–9, left valve	0·51	0·31	
,, Io.4470, right valve	0·47	0·29	
,, Io.4471, right valve	0·40	0·25	
,, Io.4472, left valve	0·36	0·26	
,, Io.4473, carapace	0·47	0·29	0·26
,, Io.4474–6, right valve	0·53	0·31	
,, Io.4477, carapace	0·50	0·26	0·26
,, Io.4478–9, left valve	0·46	0·29	

Remarks. Schizocythere aculeata (Bonnema 1941) (see Herrig 1966) is externally very close to *A. geniculata* in both ornament and carapace outline. The geniculate ridge of *S. aculeata* has an antero-dorsal projection in its anterior half in contrast to that of *A. geniculata*; the possession of a schizodont hinge is an additional distinguishing feature here and also in *Schizocythere* sp. Keij (1957), where the oblique ridge is without the geniculate bend.

Family PECTOCYTHERIDAE Hanai 1957
Genus PARAMUNSEYELLA nov.

Type species. Paramunseyella austracretacea sp. nov.

Diagnosis. Pectocytherid genus with oval/rectangular carapace, highest in anterior half, greatest length below mid-point. Anterior marginal ridge prominent, extending posteriorly on to ventro-lateral surface of carapace. Deep furrow, situated inside this ridge, follows parallel course, becoming weaker posteriorly. Carapace convex with broad flattened area below anterior cardinal angle. Postero-ventral margin denticulate. Weak, shallow depression in mid-dorsal region of each valve. Shell surface ornamented. Hinge with smooth terminal and medial elements, postero-medial tooth present. Duplicature broad with distinct incurving of inner margin. Radial pore canals few, long, slightly curved.

Remarks. Paramunseyella gen. nov. is close to *Premunseyella* gen. nov., although the latter is more quadrate in outline with a distinct postero-dorsal projection and a domed dorsal margin in the anterior half. Both genera are considered to be closely related and to occupy an ancestral position within the family with regard to the Recent genera *Pectocythere* Hanai (1957) and *Munseyella* Bold (1957), differing from these in the possession of a modified pentodont hinge and a much smaller anterior vestibule. Externally the morphological characters of the carapace are very close for all these genera. Unfortunately the muscle scars of *Paramunseyella* and *Premunseyella* are unknown at the present time so that a comparison of these characters is not possible.

Paramunseyella austracretacea sp. nov.

Plate 10, figs. 5–13; text-figs. 18A–C

1917   *Cythere lineatopunctata* Chapman and Sherborn; Chapman, p. 54, pl. 14, fig. 9.

Diagnosis. Paramunseyella with coarsely pitted shell surface. Areas between rows of pits upraised to form low ridges in antero-dorsal and postero-dorsal regions.

Holotype. Io.4498, right valve, Toolonga Calcilutite, Sample 4 (Santonian).
Paratypes. Io.4499–4505, 7 specimens from Sample 4.
Other material. 8 specimens from Sample 4.

Description. Carapace subrectangular with broadly rounded anterior margin and narrow posterior end, line of greatest length passing below mid-height. In dorsal

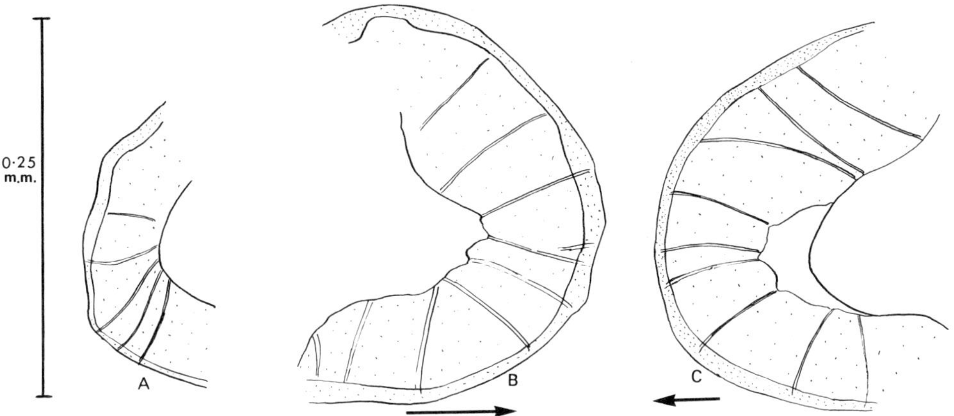

TEXT-FIGS. 18A–C. *Paramunseyella austracretacea* gen. et sp. nov. A, B, posterior and anterior duplicatures with radial pore canals, antero-median vestibule lacking because of damage, left valve, paratype Io.4504; C, anterior duplicature showing radial pore canals and median vestibule, right valve, paratype Io.4505.

view, carapace broadly convex with distinct anterior border. Region below slightly swollen anterior cardinal angle flattened, producing broad, plain-like area bounded in front by furrow and marginal ridge. Both furrow and ridge follow parallel course around anterior margin, extending on to ventral surface to die out posteriorly by curving upwards on to postero-lateral surface. Marginal denticles or spines clustered around posterior end. Shell surface strongly pitted with oblique ridges developed in antero-dorsal and postero-dorsal regions. Left valve slightly larger than right, which it overlaps along ventral margin and in region of cardinal angles.

Hinge of modified pentodont type (Hanai 1957), differing from true pentodont by having tooth developed at posterior end of median bar only. Postero-median tooth is simple, smooth tooth, not split into dorsal and ventral part as in true pentodont type. Anterior end of smooth median groove seen to widen slightly, but no evidence of antero-median tooth/socket. Terminal hinge elements smooth.

Duplicature very broad anteriorly, inner margin coinciding with line of concrescence in all parts except antero-medially, where duplicature turns in towards anterior margin, and small vestibule developed (text-fig. 18c). Presence of vestibule dependent on state of preservation. 9 long, slightly curved anterior radial pore canals and 5 posterior radial pore canals present (text-figs. 18A–C).

Muscle scars so far not identified.

Dimensions (in mm)

	Length	Height	Width
Holotype, Io.4498, right valve	0·46	0·26	
Paratype, Io.4499, right valve	0·48	0·28	
,, Io.4500, left valve	0·45	0·28	
,, Io.4501, left valve	0·48	0·28	
,, Io.4502, right valve	0·44	0·26	
,, Io.4503, carapace	0·45	0·27	0·22
,, Io.4504, left valve	0·45	0·27	
,, Io.4505, right valve	0·46	0·27	

Remarks. P. austracretacea internally closely resembles *Premunseyella ornata* sp. nov., although the latter differs externally by possessing a postero-dorsal projection, 2 rather weak dorso-median 'sulci' and a quadrate carapace.

Cythere lineatopunctata from the Gingin Chalk (Chapman 1917) is conspecific and placed in synonymy. *Cythere lineatopunctata* Chapman and Sherborn (1893) from the Gault of England is not conspecific. Chapman's specimen of *C. lineato-punctata*, CPC.7139, shows a few slight morphological variations when compared with *P. austracretacea*, for example, a more obliquely angled antero-dorsal slope, but this is here regarded as population variation.

Genus PREMUNSEYELLA nov.

Type species. Premunseyella ornata sp. nov.

Diagnosis. Pectocytherid genus with quadrate carapace, noticeably arched in region of anterior cardinal angle. Postero-dorsal slope steeply angled, almost vertical; extreme posterior end close to ventral margin. Marginal and submarginal spines at posterior end. Anterior end high, margin broadly rounded, thickened to form marginal ridge extending ventrally back along ventral surface, dorsally back along dorsal margin as far as postero-dorsal projection, with which it fuses. Deep marginal furrow runs parallel to anterior and ventral course of this ridge. Postero-dorsal part of carapace with angled ridge projecting above dorsal margin. Shell surface coarsely ornamented; 2 dorso-median furrows present. Hinge with smooth terminal and medial elements and postero-medial tooth. Duplicature broad with distinct incurving of inner margin; anterior vestibule not strongly developed. Radial pore canals few, long, slightly curved, some branching.

Remarks. As for *Paramunseyella*, the muscle scar pattern is not known, and although in many respects close to that genus, *Premunseyella* may be distinguished by its more quadrate carapace outline, postero-dorsal projection, and the presence of 2 dorso-median furrows. *Paramunseyella* appears to have only a single sulcus.

Premunseyella is very close to the Palaeocene and Recent genus *Munseyella* Bold (1957) with respect to ornamentation and carapace outline. In particular, 2 species described by Triebel (1957) from the Pleistocene of California (*Munseyella morrisi* and *M. similis*) possess the 2 weak dorso-median furrows found in *P. ornata*. The well-developed anterior vestibule of *Munseyella* is not, however, found in this genus and the hinge differs from the true pentodont, the antero-median tooth of the left valve being absent and the hinge bar smooth.

Although the muscle scar pattern is not at present known for *Premunseyella*, the close similarity of this genus to *Munseyella* suggests a phylogenetic relationship.

Premunseyella ornata sp. nov.

Plate 11, figs. 1–4, 7–9, 11; text-figs. 19A, B

Diagnosis. Premunseyella with prominent anterior marginal ridge continuous along dorsal margin to angled postero-dorsal projection and ventrally along ventral surface, turning upwards posteriorly on to lateral surface to fuse with postero-dorsal projection. Shell surface coarsely pitted with 2 weak dorso-median furrows. Male dimorph elongate; female quadrate in outline.

Holotype. Io.4506, male left valve, Toolonga Calcilutite, Sample 4 (Santonian).

Paratypes. Io.4507–4513, 7 specimens, Sample 4.

Description. Carapace small, quadrate in female dimorph; elongate in male. Anterior end broadly rounded; posterior end narrower with almost vertically angled postero-dorsal slope in female; postero-dorsal slope more obliquely angled in male. Dorsal margin typically arched in anterior two-thirds of carapace length. Thickened anterior marginal rim extends back along dorsal and ventral margins (text-figs. 19A, B) to unite at postero-dorsal projection, ventral extension posteriorly turning up to extend

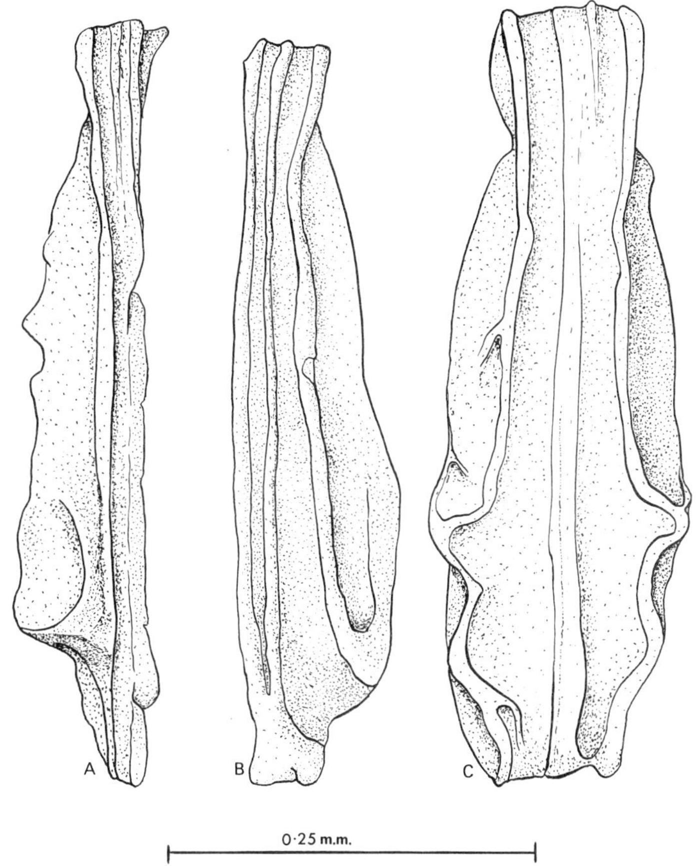

TEXT-FIGS. 19A–C. A, B, *Premunseyella ornata* gen. et sp. nov., A, B, dorsal and ventral views of left valve; note parallel ridges on ventral surface, male, holotype Io.4506. C, *Premunseyella imperfecta* gen. et sp. nov., ventral view of complete carapace to show irregular course of ventral ribbing, male, holotype Io.4514.

across lateral surface to fuse with lower surface of triangular shaped postero-dorsal projection. Lateral surface coarsely pitted, with deep marginal furrow running parallel to anterior and ventral parts of marginal ridge. 2 shallow dorso-median 'sulci' clearly visible in female dimorph; posterior sulcus is more deeply developed. Neither sulcus reflected internally by ridge, therefore not considered true sulci, but ornamental furrows. Small marginal spines at extreme posterior end of each valve. Outside anterior marginal ridge, second marginal ridge extends around anterior margin to die out dorsally at anterior cardinal angle (text-fig. 19A) but ventrally extends along ventral margin to fuse with posterior end (text-fig. 19B).

Internally, hinge of modified pentodont type. All hinge elements smooth, and terminal sockets open to interior of valve. Median bar, instead of having single terminal tooth at each end (pentodont type), has only single postero-median tooth. Not clear whether anterior part of median bar thickens slightly, but median groove appears to widen towards anterior end (Pl. 11, fig. 11).

Inner margin and line of concrescence coincide, although single specimen (Io.4513) has narrow vestibule within median incurvature of anterior duplicature. Anteriorly, duplicature very broad; few radial pore canals observed; precise number, together with information as to whether canals are single or branching, not yet determined.

Dimensions (in mm)

	Length	*Height*
Holotype, Io.4506, male left valve	0·52	0·30
Paratype, Io.4507, male right valve	0·49	0·26
,, Io.4508, male right valve	0·50	0·29
,, Io.4509, female left valve	0·47	0·30
,, Io.4510, female left valve	0·47	0·29
,, Io.4511, male right valve	0·50	0·26
,, Io.4512, female left valve	0·49	0·31
,, Io.4513, female right valve	0·47	0·28

Remarks. P. ornata is morphologically similar to such species of *Munseyella* as *M. similis* Triebel (1957) and *M. morrisi* Triebel (1957), although both differ from *P. ornata* in the generic characteristics of hinge and vestibule and also by lacking the strong surface pitting. The 2 dorso-median furrows of *P. ornata* are also present in *M. similis* and *M. morrisi.*

Premunseyella imperfecta sp. nov.

Plate 9, figs. 5–7; Plate 11, figs. 5, 6, 10; text-figs. 19C, 20

Diagnosis. Premunseyella with anterior marginal ridge extending back along ventral surface, turning outwards in posterior third to fuse with short ridge which describes an irregular course to posterior end. Thickened dorsal marginal ridge fuses with postero-dorsal projection. Carapace postero-ventrally indented; shell surface coarsely pitted.

Holotype. Io.4514, male carapace, Korojon Calcarenite, Sample 2 (Campanian).

Paratypes. Io.4515–4520. 6 specimens, 3 from Sample 2 and 3 from Sample 1, Korojon Calcarenite (Campanian).

Description. Carapace sub-rectangular without any noticeable dimorphism, although rather slender holotype could be male. Anterior end high, broadly rounded; posterior end narrower, truncated, with both marginal and submarginal denticles along postero-ventral border. Dorsal margin straight to very slightly convex, sloping posteriorly. Anterior cardinal angle broadly rounded in left valve, more angular in right, with concave antero-dorsal slope. In complete carapace, anterior cardinal angle of left valve projects strongly above right, whilst right valve projects above left along posterior part of dorsal margin. Left valve overlaps right along ventral margin.

Anterior margin thickened to form marginal rim, which dorsally passes into thickened dorsal edge of valve; this thickened ridge extends back into upstanding postero-dorsal projection. Ventrally, anterior marginal ridge extends back along ventral surface (text-fig. 19C) to turn outwards in posterior third and unite with ventro-lateral ridge which immediately turns inwards to follow somewhat sinuous course to posterior margin. In front of point of fusion of these 2 ridges, ventro-lateral ridge runs obliquely dorso-anteriorly on to lateral flank of carapace. Deep furrow inside marginal ridge runs parallel to its anterior and ventral course. Lateral surface only coarsely pitted, with dorso-median sulcation barely discernible. Carapace has prominent postero-dorsal projection and less strongly developed projection where ventral and ventro-lateral ridges fuse. Carapace between these 2 projections slightly indented.

Internally, hinge of modified pentodont type, all elements smooth. Terminal sockets of left valve open to valve interior, and smooth hinge bar has tooth-like projection at its posterior end.

Inner margin and line of concrescence appear to coincide; no definite indication of anterior vestibule observed. Duplicature anteriorly very broad; 9 anterior radial pore canals, slightly sinuous with some branching, present, together with 6 straight posterior canals and 3 ventral canals (text-fig. 20).

Muscle scars not observed.

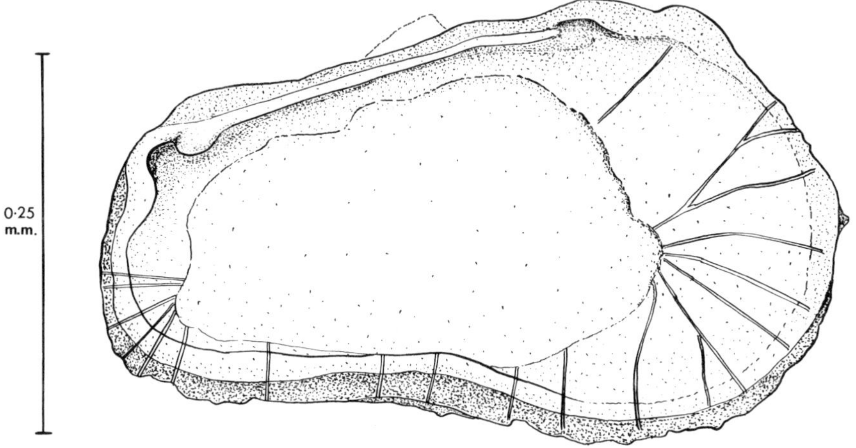

TEXT-FIG. 20. *Premunseyella imperfecta* gen. et sp. nov., internal view to show hinge, duplicature, and radial pore canals, female left valve, paratype Io.4516.

Dimensions (in mm)

	Length	Height	Width
Holotype, Io.4514, male carapace	0·52	0·29	0·18
Paratype, Io.4515, female carapace	0·51	0·27	0·21
„ Io.4516, female left valve	0·51	0·29	
„ Io.4518, female right valve	0·49	0·27	
„ Io.4519, male right valve	0·50	0·27	
„ Io.4520, female carapace	0·54	0·30	0·21

Remarks. P. imperfecta may readily be distinguished from *P. ornata* by the ventral ridge fusing with a short ventro-lateral ridge rather than with the postero-dorsal projection (text-figs. 19B, C). The postero-dorsal projection of *P. ornata* is much more strongly developed and this species does not have the postero-ventral indentation of the carapace that is present in *P. imperfecta*. Ornamentally the 2 species may also be distinguished by the presence of 2 dorso-median furrows in *P. ornata*.

Family CYTHERIDEIDAE Sars 1925
Subfamily KRITHINAE Mandelstam 1958
Genus KRITHE Brady, Crosskey and Robertson 1874
Krithe sp.

Plate 3, fig. 9; Plate 5, figs. 1, 2

Remarks. 3 single valves representing 2 distinct species have been obtained from the Toolonga Calcilutite, Sample 3 (Campanian). Only 1 of these, Io.4447 (Pl. 3, fig. 9) has the characteristic posterior indentation of the genus. The other 2 valves are much larger and possess a more angular posterior outline. No further representatives of this genus have been found in the other samples examined.

Dimensions (in mm)

	Length	Height
Io.4447, left valve	0·66	0·32
Io.4448, left valve	0·93	0·47
Io.4449, right valve	0·88	0·40

Subfamily CYTHERIDEINAE Sars 1925
Genus ROSTROCYTHERIDEA Dingle 1969*a*
Rostrocytheridea canaliculata sp. nov.

Plate 12, fig. 9; Plate 13, figs. 4, 6, 7, 8; text-figs. 21A–C

Diagnosis. Rostrocytheridea with deep furrow parallel to anterior margin. Turned down posterior end denticulate. Shell surface smooth.

Holotype. Io.4532, left valve, Toolonga Calcilutite, Sample 4 (Santonian).

Paratypes. Io.4533–4538. 6 specimens from Sample 4.

Other material. 8 specimens from Sample 4.

Description. Carapace oval, with line of greatest height passing through strongly humped anterior cardinal angle. Valves taper strongly towards posterior end, typically turned down and distinctly dentate. Dorsal margin slightly convex, sloping steeply posteriorly; ventral margin broadly convex.

Shell surface smooth, with few widely spaced normal pore canals; deep furrow parallel to thickened anterior margin extends from antero-ventral border to midway up antero-dorsal slope. Flange clearly visible around anterior margin, but ventrally present only as very low ridge. Flange present in all instars; marginal furrow only well developed in adults.

Internally, hinge antimerodont with 4–5 anterior teeth and 5–6 posterior teeth in right valve. Median groove coarsely loculate (Plate 13, fig. 8). Left valve has complementary dentate/loculate elements with very narrow accommodation groove.

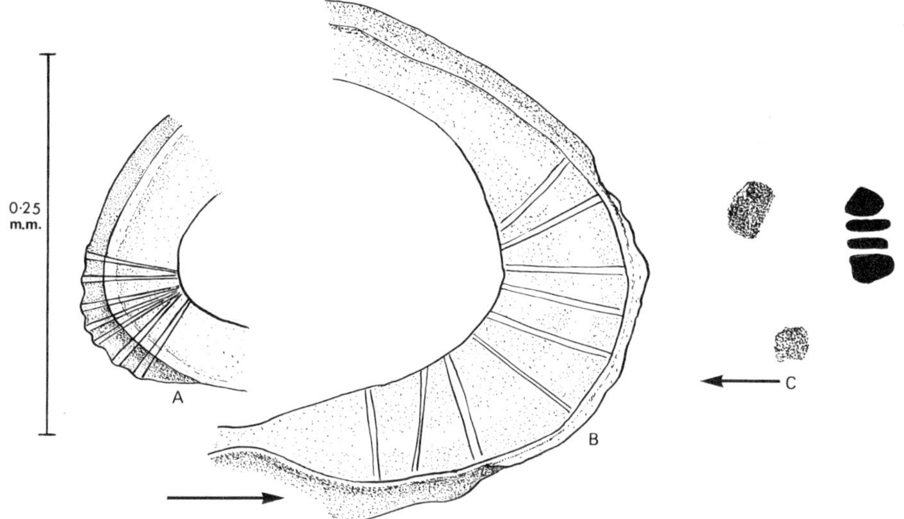

TEXT-FIGS. 21A–C. *Rostrocytheridea canaliculata* sp. nov. A, B, posterior and anterior duplicatures with radial pore canals and marginal flange, left valve, holotype Io.4532; C, muscle scars, left valve, paratype Io.4534.

Inner margin and line of concrescence coincide anteriorly, and rather broad duplicature produced. Radial pore canals long and only slightly curved, 8–9 widely spaced anteriorly, 6–7 closely grouped posteriorly (text-figs. 21A, B).

Muscle scars (text-fig. 21C) consist of vertical row of 4 adductor scars with large, oval, antero-dorsal frontal scar and smaller antero-ventral mandibular scar.

Dimensions (in mm)

	Length	Height
Holotype, Io.4532, left valve	0·60	0·36
Paratype, Io.4533, right valve	0·60	0·35
„ Io.4534, left valve	0·54	0·33
„ Io.4535, left valve	0·54	0·32
„ Io.4536, right valve	0·60	0·33
„ Io.4537, right valve	0·52	0·32
„ Io.4538, left valve	0·54	0·32

Remarks. R. canaliculata differs from *R. chapmani* Dingle (1969*a*) by possessing a smaller number of anterior radial pore canals and lacking the narrow anterior vestibule. The anterior marginal furrow is so far restricted to *R. canaliculata*. It may be necessary at a later date to separate the Australian species, distinguished from the South African *Rostrocytheridea* by the possession of a reduced number of anterior radial pore canals, either by the establishment of a new generic or a subgeneric taxon.

Subfamily EUCYTHERINAE Puri 1954
Genus EOROTUNDRACYTHERE nov.

Type species. Eorotundracythere levigata sp nov.

Diagnosis. Carapace small, triangular in outline, tapering to posterior end. Anterior end high, broadly rounded, with marginal border continued into a projecting 'ear' at anterior cardinal angle of left valve. Degree of development of 'ear' variable. Central part of carapace laterally tumid with ridge-like ventro-lateral projection. Crescentic furrow situated behind sub-central swelling, weakly or well-developed. Left valve larger than right. Shell surface smooth or ornamented. Large, circular normal pore canals prominent. Hinge antimerodont; inner margin and line of concrescence coincide posteriorly, may diverge slightly anteriorly to produce narrow vestibule. Approximately 11–14 anterior radial pore canals. Well-developed anterior marginal flange, expanded antero-ventrally. Muscle scars consist of row of 4 adductor scars with V-shaped frontal scar. Sexual dimorphism may be present.

Remarks. Eorotundracythere possesses a strongly developed antimerodont hinge and in this respect resembles the Recent species of *Eucythere* described by Hornibrook (1952) from New Zealand. Because *Eucythere* s.s. possesses a rather weak lophodont hinge, Mandelstam (1958) erected the genus *Rotundracythere* for Hornibrook's species with *Eucythere rotunda* Hornibrook as the type. The truncated posterior outline of *E. rotunda* is unlike the more typical eucytherine tapered posterior outline of the other New Zealand species which possess the antimerodont hinge.

Eorotundracythere is regarded as occupying an ancestral position to *Rotundracythere*, from which it is distinguished by the tapered posterior outline, possession of a distinct 'ear' and the almost complete absence of a vestibule.

So far, 2 distinct species of *Eorotundracythere* have been identified: *E. levigata* sp. nov. (Campanian) and *E. compta* sp. nov. (Santonian). A single valve from the Santonian, with nodose ornament, may represent a third species.

Euryitycythere Oertli (1959), a small genus from the Lower Cretaceous of Germany and France, has some similarities with *Eorotundracythere* but may be distinguished easily in dorsal outline by the projecting anterior and posterior margins and the presence of an eye node. Internally this genus also differs by possessing a considerably increased number of anterior radial pore canals.

Eorotundracythere levigata sp. nov.

Plate 14, figs. 4–8; text-figs. 22A, B, 23

Diagnosis. Eorotundracythere with unornamented shell surface. Anterior cardinal angle with distinct or reduced 'ear'.

Holotype. Io.4539, carapace from the Korojon Calcarenite, Sample 1 (Campanian).

Paratypes. Io.4540–4547, 8 specimens, 4 from Sample 1, 2 from Sample 2 (also Korojon Calcarenite—Campanian), and 2 from the Toolonga Calcilutite, Sample 3 (Campanian).

Other material. 2 specimens from Sample 1, 10 from Sample 2, and 4 from Sample 3.

Description. Carapace small, triangular in side view, highest at anterior end and tapering towards posterior end. Anterior marginal border extends dorsally above line of dorsal margin at anterior cardinal angle to produce a projecting 'ear'. This is

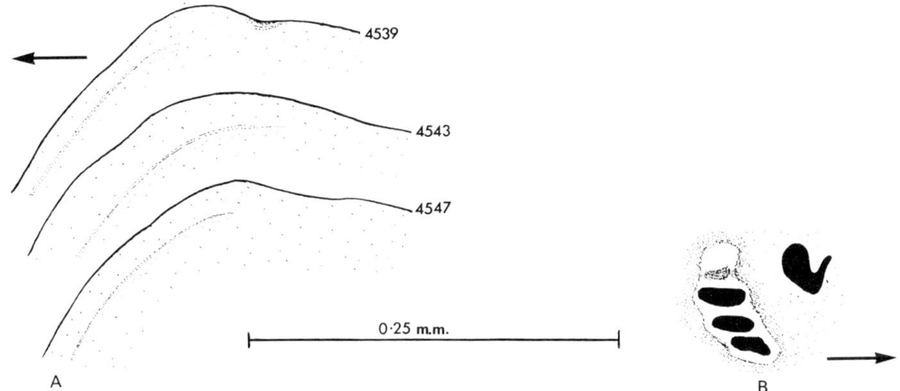

TEXT-FIGS. 22A, B. *Eorotundracythere levigata* gen. et sp. nov. A, antero-dorsal outline in 3 left valves to show variation in the species; B, muscle scars, juvenile left valve, paratype Io.4545.

developed only in left valve when present; not all specimens possess well-developed 'ear', variations from strongly projecting to broadly rounded illustrated in text-fig. 22A. Ventral margin broadly convex, as is ventrally projected ventro-lateral margin. Dorsal margin medially convex, forming hump-like outline with cardinal angles upstanding above it.

Left valve larger than right, which it overlaps along ventral margin and strongly overreaches antero-dorsally. Shell surface smooth, although rather large normal pore canal openings produce sort of ornamentation. Some specimens possess form of pitting along ventro-lateral border.

Internally, hinge antimerodont with well-developed terminal teeth (approximately 5 anteriorly and 7 posteriorly), and strongly loculate median groove in right valve. Complementary structures in left valve; narrow, rather shallow accommodation groove above median bar.

Line of concrescence and inner margin coincide posteriorly but diverge slightly anteriorly to produce very narrow vestibule and broad duplicature. Radial pore canals long, straight, approximately 14 anteriorly and 6 posteriorly. Strongly developed flange extends around anterior margin of both valves. Flange extends back along ventral surface as prominent ridge. Muscle scars consist of vertical row of 4 elongate adductor scars with V-shaped antero-dorsal frontal scar. Dorsal muscle scar just above this group of scars (text-fig. 23).

Dimensions (in mm)

		Length	*Height*	*Width*
Holotype, Io.4539, carapace		0·50	0·35	0·27
Paratype, Io.4540, carapace		0·49	0·34	0·28
,,	Io.4541, right valve	0·48	0·32	
,,	Io.4543, left valve	0·52	0·37	
,,	Io.4544, right valve	0·48	0·34	
,,	Io.4545, juvenile left valve	0·43	0·29	
,,	Io.4546, right valve	0·49	0·32	
,,	Io.4547, left valve	0·43	0·28	

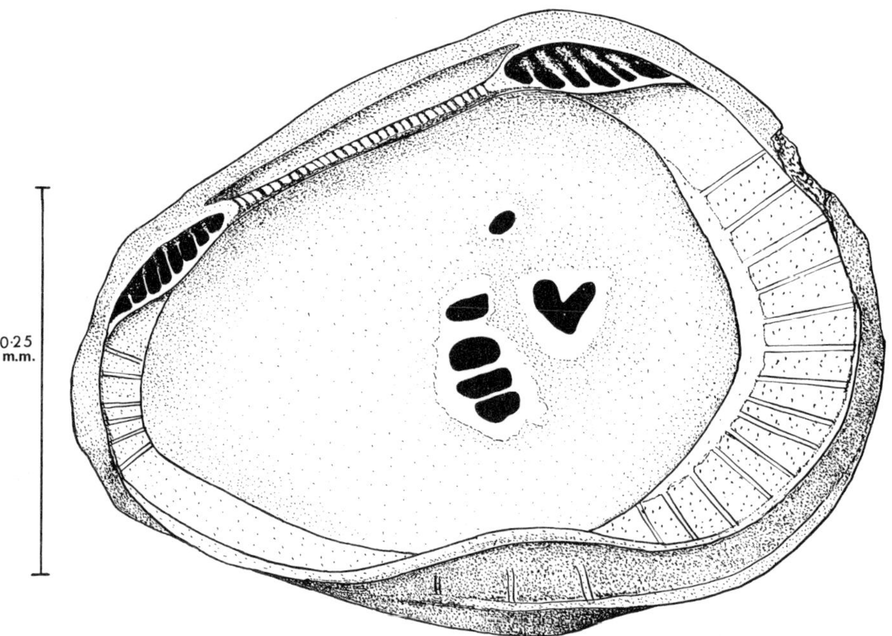

TEXT-FIG. 23. *Eorotundracythere levigata* gen. et sp. nov., internal view, left valve, to show hinge, muscle scars, duplicature with small anterior vestibule, radial pore canals, and antero-ventral flange, paratype Io.4543.

Remarks. *E. levigata* differs from species of *Eucythere* and *Rotundracythere* on characters of hinge, duplicature, and antero-dorsal 'ear'. The differences between this species and *Eorotundracythere compta* sp. nov. will be discussed under that species.

Eorotundracythere compta sp. nov.

Plate 12, figs. 3, 5, 6, 7; Plate 18, fig. 8; text-figs. 24A, B

Diagnosis. Eorotundracythere having coarsely pitted shell surface with some transverse wrinkling. Crescentic furrow behind prominent subcentral swelling. Development of lateral node in posterior half just above mid-height variable. Prominent nodes may be developed in antero-dorsal, postero-dorsal, and postero-ventral regions as well as sub-centrally. Inner margin virtually coincides with line of concrescence; 11 anterior radial pore canals. Sexual dimorphism indicated by presence of elongate males.

Holotype. Io.4548, left valve, Toolonga Calcilutite, Sample 4 (Santonian).

Paratypes. Io.4549–4555, Io.4608, 8 specimens from Sample 4.

Other material. 6 specimens from Sample 4.

Description. Carapace triangular in female dimorph, elongate in male. Anterior end high and broadly rounded; anterior cardinal angle projects prominently above dorsal outline. Posterior end tapered, narrowly rounded. Ventro-lateral margin strongly convex, almost alate, projecting below ventral surface. Prominent sub-central swelling in anterior half, with smaller swelling or node sometimes present in posterior half, short distance above and to rear; both swellings separated by deep crescentic sulcus. In 1 female left valve (Pl. 12, fig. 5), typical ornament altered so that tendency to develop nodes exaggerated, and sub-central node well developed, as also node situated above and behind; this becomes prominent postero-dorsal node. Additionally, postero-ventral node on ventro-lateral margin, and swollen ridge extends down from anterior cardinal angle to mid-height position.

In female dimorph shell surface ornamented with distinct lateral furrows extending along ventral and ventro-lateral surfaces; these appear in some specimens to curve upwards and become transverse furrows or wrinkles in antero- and postero-dorsal regions. Rather large circular pits between furrows in central area of lateral surface; towards anterior and posterior ends, pits become somewhat smaller and more numerous. In male dimorph, sub-central swelling clearly visible. Well-developed ventro-lateral ridge present and transverse ridges, instead of curving round antero- and postero-dorsal areas to join up with ventro-lateral ridges as in female, tend to

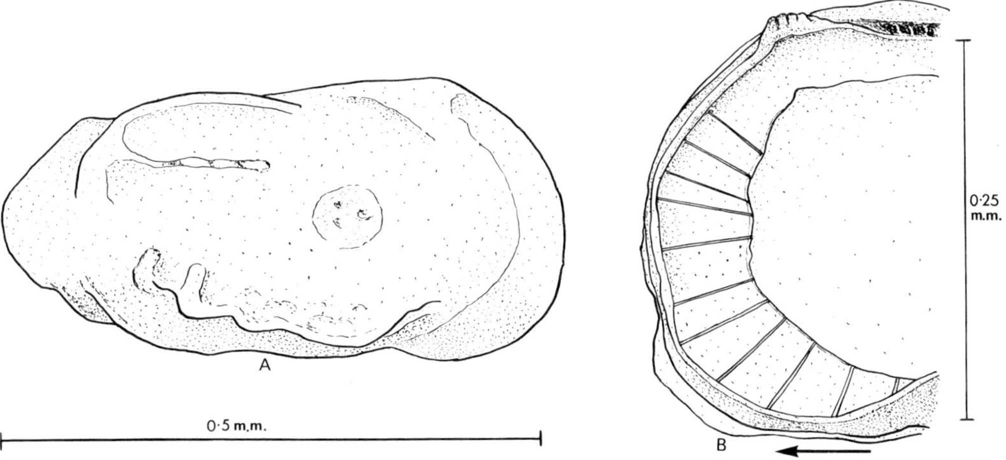

TEXT-FIGS. 24A, B. *Eorotundracythere compta* gen. et sp. nov. A, male right valve, paratype Io.4608; B, anterior duplicature with radial pore canals and anterior marginal flange, female right valve, paratype Io.4553.

D

have somewhat irregular arrangement, extending across entire lateral surface to produce reticulate appearance. Lateral pits of uniform size, unlike those of female, and smaller.

Hinge well-developed antimerodont, with approximately 5 anterior and 7 posterior teeth in right valve separated by coarsely loculate median groove. Complementary structures present in left valve; accommodation groove not present.

Selvage forms prominent ridge extending around free margin of valves. Outside selvage, broad flange extends around anterior margin and especially well-developed antero-ventrally. Flange extends back along ventral surface, where it virtually dies out to reappear posteriorly and extend around posterior margin.

Inner margin and line of concrescence virtually coincide, but anteriorly separate slightly, and narrow vestibule observed. Radial pore canals long, straight, 11 anteriorly (text-fig. 24B) and approximately 3–4 posteriorly. Only 4 oval adductor muscle scars observed; frontal scar not visible.

Dimensions (in mm)

	Length	*Height*
Holotype, Io.4548, female left valve	0·51	0·35
Paratype, Io.4549, male left valve	0·60	0·33
„ Io.4550, female right valve	0·45	0·28
„ Io.4551, female right valve	0·50	0·31
„ Io.4552, female left valve	0·50	0·32
„ Io.4553, female right valve	0·47	0·28
„ Io.4554, female right valve	0·49	0·29
„ Io.4555, female left valve	0·40	0·28
„ Io.4608, male right valve	0·56	0·29

Remarks. E. compta is easily distinguished from *E. levigata* by the surface ornamentation, deep crescentic sulcus, and more prominently developed anterior cardinal process. Sexual dimorphism has not been observed in *E. levigata*.

The appearance of a strongly nodose form of *E. compta* is interesting, and given more material may be identified as a distinct species. At the moment this form is regarded as being only a nodose variant.

Family CYTHERURIDAE Müller 1894

Remarks. Within the Cytheruridae a number of genera possess a hinge in which the median element expands at both ends and becomes more coarsely dentate/loculate. This is commonly referred to as the *Cytheropteron* hinge. Examination of this structure has revealed 4 variations of this hinge type, which is here named the *peratodont* (Greek, *paras*, end) type. The variations of this hinge are as follows:

1. *Holoperatodont* (text-fig. 25A). All the hinge elements are smooth, e.g. *Cytheropteron* (*Infracytheropteron*) Kaye (1964). The development of the median element is often such that the terminal enlargements are not obvious; a good example is shown, however, in *C. (Infracytheropteron) anotum* sp. nov. (Pl. 19, fig. 4).

2. *Hemiperatodont* (text-fig. 25B). The terminal elements are dentate/loculate whilst the median elements are smooth, e.g. *Cytherura* Sars (1866).

3. *Paraperatodont* (text-fig. 25C). The terminal elements of the hinge, as well as the terminal elements of the median bar/groove, are coarsely dentate/loculate, whilst

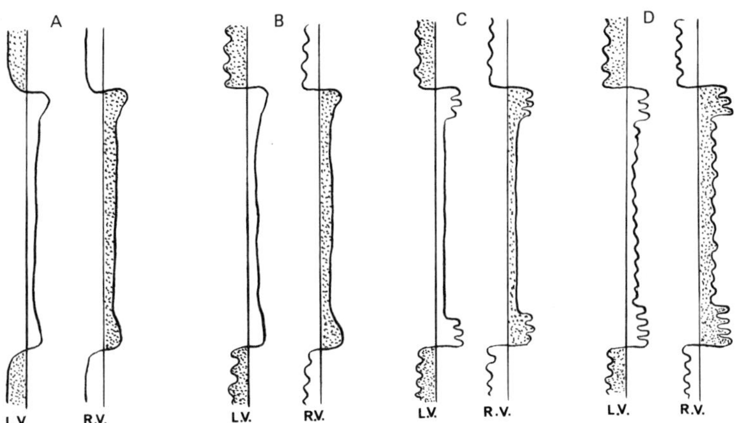

TEXT-FIGS. 25A–D. Peratodont hinge: A, Holoperatodont; B, Hemiperatodont; C, Paraperatodont; D, Artioperatodont. Dorsal view of left and right valves with sockets and grooves shaded.

the centre portion of the median bar/groove is smooth (Pl. 19, fig. 2). This type is present in both *Oculocytheropteron* gen. nov. and *Semicytherura* Wagner (1957).

4. *Artioperatodont* (text-fig. 25D). All the hinge elements are coarsely dentate/loculate in this type, which is found in *Hemicytherura* Elofson (1941), *Oculocytheropteron* gen. nov., and *Cytheropteron* (*Cytheropteron*) Sars (1866). In the latter case the degree of development of the terminal teeth of the median bar is variable, so much so that the hinge may be either artioperatodont, or approach an antimerodont condition (Pl. 19, fig. 1).

Hanai (1957, text-fig. 1) described a new hinge which also had an anterior and posterior development of the median bar. This pentodont hinge differs from the peratodont type by having an upper and a lower tooth at both ends of a crenulate median bar, the terminal elements being smooth.

Subfamily CYTHERURINAE Müller 1894
Genus CYTHERURA Sars 1866
?*Cytherura* sp.

Plate 18, fig. 6

Remarks. 1 carapace (Io.4563) from the Korojon Calcarenite, Sample 1 (Campanian). Length 0·36 mm, height 0·19 mm, width 0·17 mm.

Subfamily CYTHEROPTERINAE Hanai 1957
Genus CYTHEROPTERON Sars 1866
Subgenus CYTHEROPTERON Sars 1866
Cytheropteron (*Cytheropteron*) *carinoalatum* sp. nov.

Plate 17, figs. 10, 11; Plate 18, figs. 1–5; Plate 19, fig. 1; text-figs. 26A, B, 27A, B

Diagnosis. Large species (length 0·78–0·93 mm) of subgenus *Cytheropteron* with high, broadly convex dorsal margin; straight hinge, well-developed alae. Anterior and posterior ends narrow; greatest length of carapace through mid-point. Shell surface either smooth or with transverse wrinkles. Alae possess 3 parallel ridges, outermost forming thickened alar margin. Ventral surface with 3 or 4 longitudinal ridges.

Holotype. Io.4565, left valve, Korojon Calcarenite, Sample 2 (Campanian).

Paratypes. Io.4566–4573. 1 right valve, Sample 2; 4 specimens from the Toolonga Calcilutite, Sample 3 (Campanian); 2 specimens from the Korojon Calcarenite, Sample 1 (Campanian), and 1 specimen from the Toolonga Calcilutite, Sample 4 (Santonian).

Other material. 10 specimens from Sample 1, 9 from Sample 2, 10 from Sample 3, and 4 from Sample 4.

Description. Carapace of large size, strongly convex in dorsal and ventral views. Laterally, dorsal margin broadly convex, extending from narrow anterior end to posterior end with only slight concavity to indicate presence of postero-dorsal slope. Ventro-lateral margin distinctly alate, projecting both laterally and ventrally.

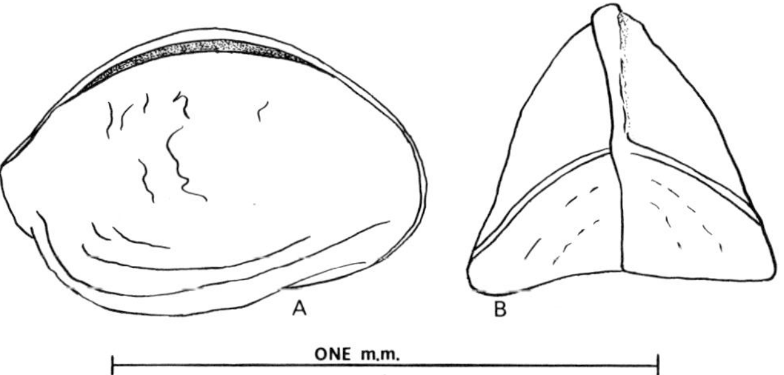

TEXT-FIGS. 26A, B. *Cytheropteron* (*Cytheropteron*) *carinoalatum* sp. nov. Right side and anterior end view, complete carapace, paratype Io.4570.

Ventral margin convex, with shallow antero-median concavity. Posterior end triangular and slightly caudate. Ventral surface ornamented by 3 or 4 longitudinal ridges and prominent normal pore canal openings. Laterally, shell surface may be smooth or possess transverse wrinkles. Both variants occur together in same sample. Most specimens have 2 or 3 distinct ridges parallel to alar margin, outermost forming edge of ala. 1 completely smooth specimen observed. Thickened anterior margin of left valve projects dorsally above smaller right valve. Normal pore canals large and widely spaced over lateral surface, but closely grouped along alar margin, where they take on appearance of radial pore canals.

Internally, hinge coarsely artioperatodont with 6 anterior and 7 posterior teeth in right valve. Loculate median groove becomes more coarsely loculate at both ends and overhung by dorsal margin of valve (Pl. 18, fig. 2). In left valve (Pl. 17, fig. 10) dentate median bar projects beyond line of dorsal margin, and distinct accommodation groove present.

Inner margin and line of concrescence do not coincide antero-ventrally, and narrow vestibule is developed; elsewhere the 2 coincide to produce broad duplicature. Anteriorly, 14 simple, slightly curved radial pore canals (text-fig. 27A); 6 grouped in pairs posteriorly (text-fig. 27B). Posterior canals bifurcate at their outer extremity.

Outside selvage, distinct flange extends around anterior and antero-ventral margins, and around posterior and postero-ventral margins.

Muscle scars (Pl. 18, fig. 5) consist of 4 elongate adductor scars in vertical row with V-shaped frontal scar situated slightly below antero-median position.

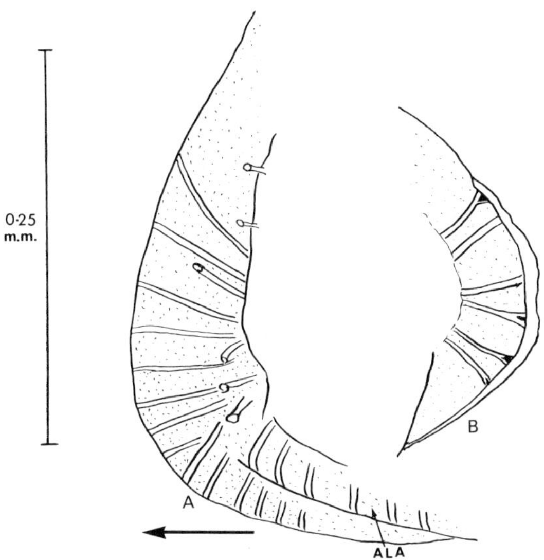

TEXT-FIGS. 27A, B. *Cytheropteron* (*Cytheropteron*) *carino-alatum* sp. nov. Anterior and posterior duplicatures with radial pore canals; note terminal branching in posterior canals and, anteriorly, false radial canals and normal pore canals in ventral ala; external views, left valve, paratype Io.4567.

Dimensions (in mm)

	Length	Height	Width
Holotype, Io.4565, left valve	0·78	0·52	
Paratype, Io.4566, right valve	0·84	0·45	
,, Io.4567, left valve	0·81	0·56	
,, Io.4568, carapace	0·72	0·50	0·51
,, Io.4569, right valve	0·84	0·48	
,, Io.4570, carapace	0·77	0·51	0·53
,, Io.4571, left valve	0·85	0·59	
,, Io.4572, left valve	0·93	0·68	
,, Io.4573, carapace	0·79	0·49	0·58

Remarks. *C. carinoalatum* has a distinct carapace outline and a variable ornament. The degree of ornamentation illustrated for the holotype and one of the paratypes (Pl. 17, fig. 11; Pl. 18, fig. 1) varies within individuals and appears, in part, to be related to an increase in size. The largest specimen (Io.4572), for example, is perfectly smooth and even lacks the alar ridges which are present in otherwise smooth specimens. All the variants occur together in the same sample.

In carapace outline the Recent *Cytheropteron fenestratum* Brady (1880) is similar to *C. carinoalatum* but differs in possessing dentate terminal margins, a row of pits parallel to the alar margin and a more strongly tapered posterior end.

Subgenus INFRACYTHEROPTERON Kaye 1964

Remarks. The identification of a new species of *Infracytheropteron* confirms that the subgenus is represented by small (length 0·32–0·49 mm), generally smooth species, rather tumid in outline, and tapering to the posterior. Only a few (5–8) anterior radial pore canals are present and all the elements of the hinge are smooth.

Cytheropteron (Infracytheropteron) anotum sp. nov.

Plate 17, figs. 1–9; Plate 19, fig. 4

Diagnosis. Species of *Infracytheropteron* without surface ornamentation. Dorsal margin domed; carapace tumid, posteriorly acuminate.

Holotype. Io.4574, carapace, Korojon Calcarenite, Sample 1 (Campanian).
Paratypes. Io.4575–4580, 6 specimens from Sample 1.
Other material. 4 specimens from Sample 1.

Description. Carapace small, rather tumid in outline with high domed dorsal margin, degree of convexity variable (see Pl. 17, figs. 1–4). Posterior cardinal angle distinct, rather acute; anterior cardinal angle indistinct. Ventral margin convex, overhung by downwardly projected and well-rounded alae. Anterior margin broadly rounded; posterior end tapering, distinctly acuminate in some species.

Shell surface smooth except for small number of normal pore canal openings which tend to produce pitted appearance. Left valve larger than right, overreaches it dorsally but overlaps it ventrally.

Hinge of holoperatodont type (text-fig. 25A); only clearly observed in left valve. Terminal sockets open to interior of valve (Pl. 17, fig. 7) and absence of any loculation indicates teeth not present in right valve. Median bar, as described by Kaye (1964), smooth, but as seen from illustrations (Pl. 17, fig. 8; Pl. 19, fig. 4) expands terminally. This feature not described by Kaye in his original or subsequent species descriptions, although found to be present in specimens of *Cytheropteron (Infracytheropteron) lindumensis* Kaye and Barker (1965).

Inner margin and line of concrescence not quite coincident anteriorly, and narrow vestibule produced. Radial pore canals not clearly seen; appear to be about 5 straight canals anteriorly. No muscle scars observed.

Dimensions (in mm)

	Length	Height	Width
Holotype, Io.4574, carapace	0·49	0·35	0·26
Paratype, Io.4575, left valve	0·47	0·32	
,, Io.4576, left valve	0·49	0·31	
,, Io.4577, left valve	0·49	0·32	
,, Io.4578, carapace	0·47	0·31	0·26
,, Io.4579, right valve	0·43	0·27	
,, Io.4580, juvenile left valve	0·34	0·22	

Remarks. Cytheropteron (*Infracytheropteron*) *exquisitum* Kaye (1964), the type species, is an ornate form whilst other species assigned to the subgenus are smooth. *C. (I.) anotum* differs from the latter forms in the possession of a rather deep, tumid carapace and an acuminate posterior end.

Genus OCULOCYTHEROPTERON nov.

Type species. Oculocytheropteron praenuntatum sp. nov.

Diagnosis. Cytheropterinae with well-developed eye tubercle situated just below anterior cardinal angle. Alae well developed; anterior margin of ala cuts obliquely upwards on to lateral surface, sometimes reaching anterior margin. Dorsal margin arched with steeply angled antero-dorsal slope, often concave in right valve. Hinge curved, artioperatodont or paraperatodont. Duplicature may have small anterior vestibule; radial pore canals straight, simple, few. Left valve overlaps right ventrally; right valve overreaches left dorsally. Muscle scars an oblique row of 4 oval adductor scars with oval antero-dorsal frontal scar.

Remarks. 2 variations of the peratodont hinge have been observed in this genus, but so far only the Upper Cretaceous *O. praenuntatum* possesses the paraperatodont type. The Recent species listed below all possess the artioperatodont hinge. Additional fossil species will have to be identified before a decision can be made on the desirability of subdividing the genus on the hinge type.

Oculocytheropteron brings together those species previously placed in *Cytheropteron* which possess a distinct eye tubercle, an oval frontal scar, and an ala in which the anterior margin cuts upwards across the lateral surface. *Cytheropteron* s.s. is a blind genus and was diagnosed as such by Sars (1866). The geological range of *Oculocytheropteron* is from the Santonian to Recent, and appears, if limited to those listed below, to be both geologically and at the present time restricted to the Southern Hemisphere.

> *Oculocytheropteron praenuntatum* sp. nov., Santonian, Western Australia.
> *Cytheropteron assimile* Brady 1880, Recent, South Indian Ocean.
> *C. acutangulum* Hornibrook 1952, Oligocene–Recent, New Zealand.
> *C. confusum* Hornibrook 1952, Oligocene–Recent, New Zealand.
> *C. dividentum* Hornibrook 1952, Oligocene–Recent, New Zealand.
> *C. fornix* Hornibrook 1952, Eocene–Recent, New Zealand.
> *C. improbum* Hornibrook 1952, Eocene–Recent, New Zealand.
> *C. gaussi* Müller 1908 (see Neale 1967), Recent, Antarctica.

Because they do not possess the upward sweep of the anterior edge to the ala, the following are considered to form a distinct subgroup:

> *Cytheropteron curvicaudum* Hornibrook 1952, Recent, New Zealand.
> *C. terecaudum* Hornibrook 1952, Recent, New Zealand.
> *C. vertex* Hornibrook 1952, Recent, New Zealand.

One Northern Hemisphere eyed *Cytheropteron*, *C. nodosum* Brady 1868, is here tentatively included in *Oculocytheropteron*. Recorded from the Pleistocene and Recent of northern Europe, the eye tubercle of this species is not so well developed and internally does not have such a distinct opening as in the southern species. Also the anterior vestibule is much larger and the frontal muscle scar, instead of being oval, is a long, narrow, obliquely sloping scar with a small circular scar situated above. These differences may be insignificant; if so, the inclusion of *C. nodosum* in *Oculocytheropteron* gives the genus a northern and southern distribution. Should the differences mentioned above be strengthened by the identification of other northern species, then it may be preferable to separate the 'eyed' cytheropterons into two distinct groups.

Oculocytheropteron praenuntatum sp. nov.

Plate 16, figs. 1–12; Plate 19, figs. 2, 3; text-figs. 28A–C

Diagnosis. Oculocytheropteron with paraperatodont hinge; coarsely pitted, almost reticulate surface ornament with fine pits along anterior marginal border. Alae with thickened marginal keel, low swelling at base of anterior alar margin. Antero-dorsal slope concave in right valve. Low dorsal ridge runs parallel to dorsal margin, extending from eye tubercle to posterior region.

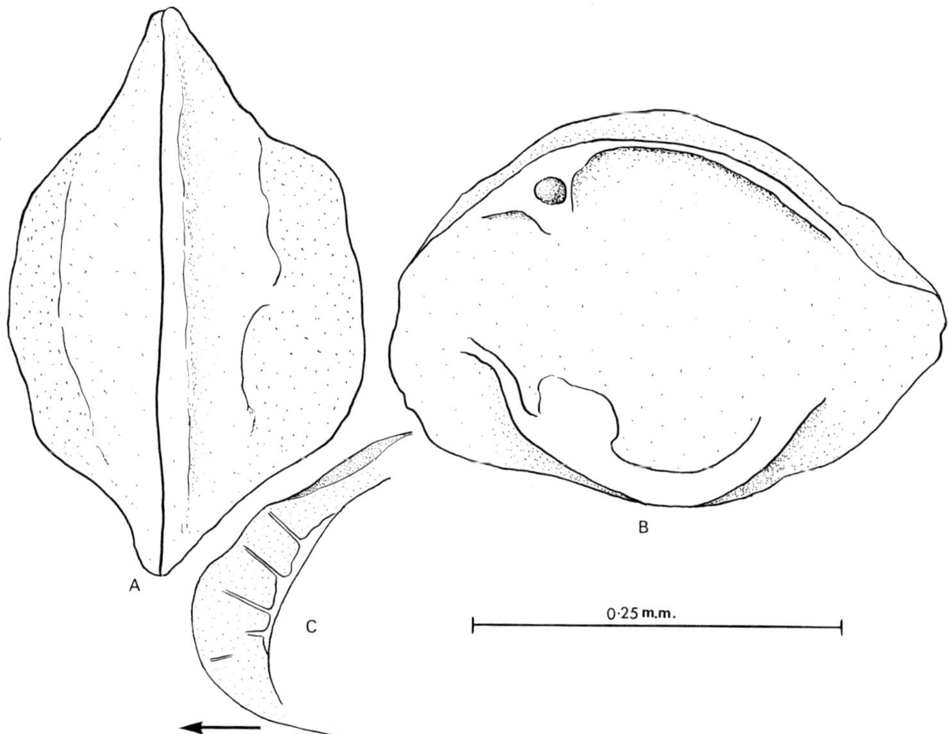

TEXT-FIGS. 28A–C. *Oculocytheropteron praenuntatum* gen. et sp. nov. A, B, dorsal and left sides, complete carapace (note prominent eye tubercle, dorsally projecting right valve, and ventral ala with anterior swelling on alar margin), paratype Io.4583; C, internal view, right valve duplicature to show anterior radial pore canals and vestibule, paratype Io.4585.

Holotype. Io.4581, left valve, Toolonga Calcilutite, Sample 3 (Campanian).

Paratypes. Io.4582–4589; 1 specimen from Sample 3; 3 specimens from the Korojon Calcarenite, Sample 1 (Campanian), and 2 specimens from the Toolonga Calcilutite, Sample 4 (Santonian).

Other material. 1 specimen, Sample 1; a juvenile specimen from the Korojon Calcarenite, Sample 2 (Campanian), and 1 juvenile specimen, Sample 4.

Description. Carapace ovoid, tapering towards posterior end. Anterior end narrow in right valve with oblique, concave, antero-dorsal slope; in left valve broader, with convex antero-dorsal slope. Posterior end triangular, slightly upturned. Dorsal margin broadly convex with sharply angled posterior cardinal angle. Prominent, round, eye tubercle just below anterior cardinal angle. Internal ocular pit beneath anterior hinge element (Pl. 16, figs. 8, 10) reveals convex inner surface of eye 'lens'. Backwardly projected alae well developed and thickened at extreme tip and along lower part of anterior alar margin. In majority of specimens low swelling at base of anterior alar margin where it joins lateral surface of carapace. This margin cuts obliquely upwards across lateral surface to extend almost to antero-dorsal margin.

Shell surface rather coarsely pitted, almost reticulate towards centre of valve; towards anterior end pits decrease in size to become very small in region of anterior margin. In well-preserved specimens, reticulate pattern of very fine striae covers caudal region. Low dorsal ridge extending from eye tubercle runs parallel to dorsal margin as far as posterior cardinal angle in some specimens, but may extend further into posterior region in others. Ventral surface has approximately 4 parallel, low ridges, which anteriorly curve upwards on to lateral surface below anterior alar margin.

Left valve overlaps right along central part of ventral margin; dorsally, right valve projects above left (text-figs. 28A, B).

Internally, paraperatodont hinge (text-fig. 25C) with curved median hinge bar of left valve (Pl. 19, fig. 2) quite smooth. Terminally, approximately 2 anterior and 2 posterior teeth in right valve (Pl. 16, figs. 9, 11, 12).

Muscle scars consist of vertical row of 4 oval adductor scars with oval antero-dorsal frontal scar (Pl. 16, fig. 10; Pl. 19, fig. 10).

Inner margin and line of concrescence virtually coincide anteriorly although very narrow vestibule present (text-fig. 28C), with only very few straight radial pore canals crossing broad duplicature anteriorly. Precise number of radial pore canals not observed.

Dimensions (in mm)

	Length	Height	Width
Holotype, Io.4581, left valve	0·38	0·25	
Paratype, Io.4582, left valve	0·37	0·21	
,, Io.4583, carapace	0·42	0·29	0·26
,, Io.4584, right valve	0·37 (broken)	0·28	
,, Io.4585, right valve	0·43	0·28	
,, Io.4586, right valve	0·42	0·26	

Remarks. *O. praenuntatum* is the only species of this genus which has a paraperatodont hinge. Ornamentally it differs from those species described by Hornibrook (1952) which have been included in the genus by the nature of the surface pitting. The dorsal ridge of *O. praenuntatum* is also present in *O. acutangulum* (Hornibrook), *O. improbum* (Hornibrook), and *O. confusum* (Hornibrook).

Genus PARACYTHERIDEA Müller 1894
?*Paracytheridea* sp.

Plate 18, fig. 7

Remarks. 1 right valve (Io.4564) from the Korojon Calcarenite. Sample 2 (Campanian); length 0·33 mm, height 0·18 mm.

Subfamily Uncertain
Genus ORTHONOTACYTHERE Alexander 1933
?*Orthonotacythere* sp.

Plate 6, fig. 1

Remarks. 1 right valve (Io.4496) with irregularly distributed tubercles on the shell surface. Hinge antimerodont. Specimen found in the Korojon Calcarenite, Sample 1 (Campanian); length 0·43 mm, height 0·22 mm. The size of this specimen may indicate that it is a juvenile instar. Until additional material becomes available the assignment with query to *Orthonotacythere* is considered to be satisfactory.

Genus PEDICYTHERE Eagar 1965
Pedicythere sp.

Plate 4, figs. 3, 4, 9

Remarks. 1 carapace (Io.4497) present in the Toolonga Calcilutite, Sample 3 (Campanian); length 0·38 mm, height 0·17 mm, width 0·17 mm (inclusive of alae).

This new species, possessing the characteristic postero-ventral process of *Pedicythere*, is the first record of the genus outside Europe. It differs from *Pedicythere tessae* Eagar (1965) by the presence of a flattened antero-dorsal projection and a shorter ventro-lateral ala. Internal details unknown.

Family TRACHYLEBERIDIDAE Sylvester-Bradley 1948
Subfamily TRACHYLEBERIDINAE Sylvester-Bradley 1948
Genus ANEBOCYTHEREIS nov.

Type species. Anebocythereis amoena sp. nov.

Diagnosis. Trachyleberid genus with subrectangular carapace bearing distinct reticulate ornament; hexagonal pits produced having short, triangular, radial spikes extending into them from reticulae. Short stubby spines at nexus of reticulation. Small but distinct eye node below anterior cardinal angle. Anterior and postero-ventral margins dentate. Shallow median sulcus behind indistinct subcentral swelling. Hinge hemimerodont. Duplicature very narrow; radial pore canals short and straight. Muscle scars with V-shaped frontal scar.

Remarks. Anebocythereis is identified as a trachyleberid genus on carapace outline, possession of an admittedly rather poor sub-central swelling, and a V-shaped frontal scar. Like *Acanthocytheris* Howe (1963), it possesses a reticulate surface ornament coupled with superimposed spines, although these are not as well developed. Also the duplicature is rather narrow for the size of the individuals. The 2 genera differ in the more simplified nature of the hinge in *Anebocythereis* and the absence of the distinct anterior marginal border of *Acanthocythereis*. The possession of a hemimerodont hinge and a narrow duplicature could suggest a juvenile trachyleberid condition. This is not considered to be the case here, however, as the size of the individuals, the presence of juvenile instars, and the absence of any larger specimens strongly supports the contention that the larger specimens are adults. Even if the larger specimens were of pre-adult instars, at this size they should possess the adult hinge.

Anebocythereis amoena sp. nov.

Plate 13, figs. 1, 2, 5; Plate 14, figs. 1–3; Plate 15, figs. 1–3, 5, 6

Diagnosis. Anebocythereis with postero-dorsal and postero-ventro-lateral projection, especially well developed in juvenile instars where hollow spine is present.

Holotype. Left valve, Io.4556, Toolonga Calcilutite, Sample 3 (Campanian).

Paratypes. Io.4557–4562, 6 valves, Sample 3.

Other material. 9 valves, Sample 3, and 2 valves, Korojon Calcarenite, Sample 2 (Campanian).

Description. Carapace subrectangular, with broad, high, anterior margin and narrow triangular posterior end. Line of greatest height passes through anterior cardinal angle, below which is small but prominent eye tubercle. Dorsal margin straight, sloping posteriorly; ventral margin broadly convex with antero-median concavity. Postero-dorsal slope either straight or slightly concave; postero-ventral slope convex. Postero-ventral and anterior margins both dentate in adult instars, whereas in juvenile instars anterior margin nearly always non-dentate. Just anterior to mid-point carapace distinctly swollen in region of trachyleberid sub-central tubercle, with shallow median sulcus behind.

Surface strongly reticulate, with short, radial spikes extending into hexagonal pits (Pl. 15, figs. 5, 6). In anterior part of carapace, main ridges of reticulation and pits arranged in rows parallel to anterior margin; elsewhere, no apparent alignment with other valve margins. In postero-dorsal and postero-ventral-lateral regions adult instars tend to have carapace slightly projected, with short, stubby spines, especially postero-dorsally. This development is residual from juvenile condition, where strong spine is present in both regions. Spines hollow, but smaller spine bases freely scattered over carapace are apparently not. Stubby spines occur on ridges of reticulae (Pl. 14, figs. 1, 2; Pl. 15, fig. 5), terminal hollows being produced when tips of spines broken off.

Eye tubercle opens to carapace interior just below anterior hinge socket (left valve) or hinge teeth (right valve). Hinge long, rather weak; median bar/groove smooth; terminal teeth in right valve (Pl. 13, fig. 2) poorly developed. Posterior socket of left valve opens to interior of valve, whereas anterior socket within anterior end of hinge bar and shows weak loculation (Pl. 15, fig. 2).

Muscle scars consist of slightly oblique row of 4 elongate adductor scars with lowermost rather small (Pl. 15, fig. 1). Frontal scar large, V-shaped; mandibular scar oval.

Duplicature very narrow with inner margin coinciding with line of concrescence. Radial pore canals short and straight in antero-ventral region; not observed elsewhere, and precise number not known.

Dimensions (in mm)

	Length	*Height*
Holotype, Io.4556, left valve	0·88	0·54
Paratype, Io.4557, left valve	0·83	0·50
„ Io.4558, right valve	0·80	0·49
„ Io.4559, right valve	0·81	0·52
„ Io.4560, left valve	0·82	0·50
„ Io.4561, left valve	0·82	0·51
„ Io.4562, juvenile right valve	0·63	0·40

Remarks. Externally *A. amoena* closely resembles *Cythereis hostizea* Hornibrook (1952) but internally there is a considerable difference between the 2 with regard to the hinge, duplicature, and radial pore canals. The narrow duplicature, hemimerodont hinge, and impressive ornament make *A. amoena* a distinct species.

Genus COSTA Neviani 1928

Costa elongata sp. nov.

Plate 21, figs. 8, 9; text-figs. 29A–C

Diagnosis. *Costa* (s.l.) with elongate carapace in lateral view, coarsely reticulate shell ornament; rather weak, oblique median ridge; dorsal and ventro-lateral ridges most distinct in posterior half.

Holotype. Io.4628, carapace, Korojon Calcarenite, Sample 1 (Campanian).

Paratypes. Io.4629–4630, 2 right valves from Sample 1.

Description. Carapace slender, elongate in lateral view with line of greatest length passing through mid-point in left valve but passing below mid-point in right valve. Posterior end triangular, anterior end broadly rounded. Marginal denticles present around anterior margin and along postero-ventral margin. Distinct knob-like projection of posterior cardinal angle present in left valve; commonly found amongst Trachyleberididae. Eye swelling at anterior cardinal angle and from it distinct anterior marginal ridge extends around anterior margin to die out antero-ventrally. Of 3 lateral ridges, dorsal and ventro-lateral ridges are best developed in anterior half of carapace; median ridge weakly developed, oblique, with sub-central swelling situated along its length.

Shell surface further ornamented by coarse reticulation, which on ventral surface gives rise to 3 fine, longitudinal ridges (text-fig. 29B).

Left valve larger than right, overlapping it at cardinal angles and along ventral margin.

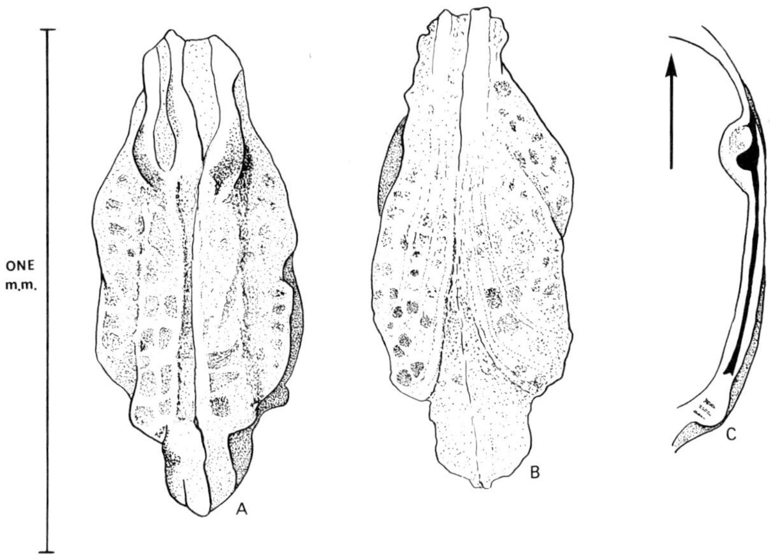

TEXT-FIGS. 29A–C. *Costa elongata* sp. nov.   A, B, dorsal and ventral views, complete
carapace, holotype Io.4628; C, right valve hinge, paratype Io.4629.

Internally, hinge amphidont, apparently holamphidont, although some weak
crenulation of posterior tooth visible (text-fig. 29C).

Duplicature broad, without vestibule, and with unknown number of radial pore
canals. Selvage well developed around free margin of right valve. Muscle scars not
observed.

Dimensions (in mm)

	Length	*Height*	*Width*
Holotype, Io.4628, carapace	0·95	0·48	0·42
Paratype, Io.4629, right valve	0·90	0·44	
„ Io.4630, right valve	0·90	0·45	

Remarks. The elongate lateral outline of *C. elongata* provides a ready feature which
distinguishes this species from previously described forms. Perhaps the anterior
marginal ridge is not as distinct as for some species of the genus, but even in the
absence of the muscle scar pattern, and number of radial pore canals, *C. elongata*
does not readily fit into any other generic unit. Because later subdivision of the
genus may become necessary the species is referred to *Costa* s.l.

Genus CURFSINA Deroo 1966
(= *Hazelina* Moos 1966 and *Repandocosta* Hazel 1967*a*)

Remarks. 3 genera, *Curfsina*, *Hazelina*, and *Repandocosta*, were independently
introduced for this small genus characterized by its postero-dorsally upturned lateral
ridge. *Curfsina* Deroo has priority over the other 2.

Curfsina levigata sp. nov.

Plate 21, figs. 5–7, 10–12; text-figs. 30A–F

Diagnosis. Curfsina with rectangular shaped carapace and smooth shell surface. Small punctae observed under high magnification only.

Holotype. Io.4631, female carapace, Korojon Calcarenite, Sample 1 (Campanian).

Paratypes. Io.4632–4634. 1 female carapace from Sample 1, and 2 valves (male) from Sample 2, Korojon Calcarenite (Campanian).

Description. Carapace rectangular in lateral view, narrowing slightly towards posterior end. Line of greatest length passes through mid-point in left valve, but tends to lie below this in right valve. Distinct eye swelling just below anterior cardinal angle of left valve; not so well developed in right valve; this true for both male and female dimorphs.

Of 3 lateral ridges, dorsal ridge is somewhat undulating in outline; ventro-lateral ridge distinct and median ridge short and oblique, extending back from sub-central swelling to turn upwards posteriorly and unite with posterior end of dorsal ridge. Depending on state of preservation, anterior and postero-ventral margins may or may not be dentate. Shell surface apparently quite smooth, but under high magnification small, circular punctae visible (Pl. 21, fig. 10). Pits illustrated in Plate 21, fig. 11 are erosional solution pits not related to pits mentioned above. Left valve posterior cardinal angle has knob-like projection. Left valve larger than right, which it overlaps at cardinal angles and along ventral margin.

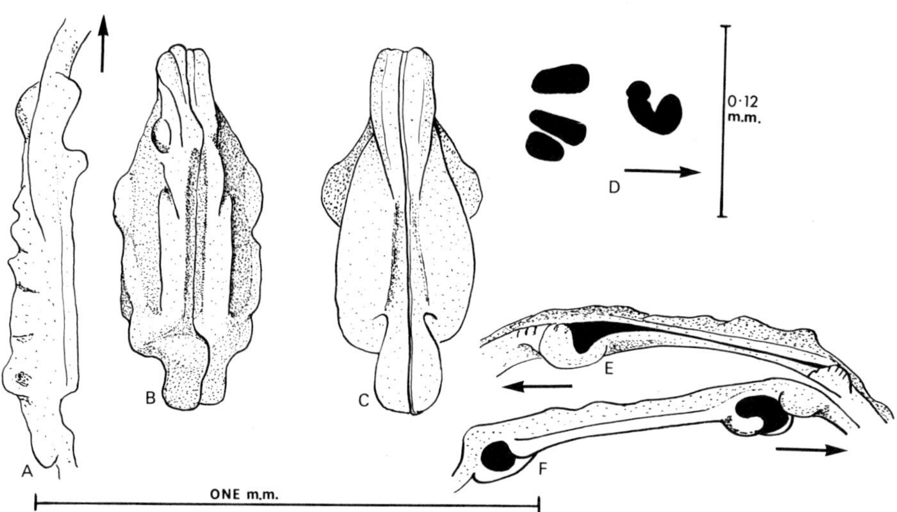

TEXT-FIGS. 30A–F. *Curfsina levigata* sp. nov. A, D, F, dorsal view of left valve hinge; muscle scars with posterior arm of the frontal scar constricted and formed by the fusion of two scars; lateral view of left valve hinge, male left valve, paratype Io.4633; B, C, dorsal and ventral views of female carapace, holotype Io.4631; E, lateral view, right valve hinge, male, paratype Io.4634.

Internally, hinge amphidont, with weak crenulations marking otherwise smooth anterior and posterior teeth. In front of anterior tooth, 3 crenulations apparent (text-fig. 30E), possibly serving to articulate with swollen projection in left valve just in front of anterior socket (text-fig. 30F). Projection similar to anterior tooth in genus *Idiocythere* Triebel (1958).

Inner margin and line of concrescence coincide to produce moderately broad duplicature with 30–40 radial pore canals anteriorly, only poorly observed but apparently simple canals.

Muscle scars (text-fig. 30D) only partially preserved in present material; only 3 adductor scars seen. Frontal scar V-shaped, apparently formed by combination of 2 muscle scar spots.

Dimensions (in mm)

	Length	*Height*	*Width*
Holotype, Io.4631, female carapace	0·72	0·40	0·32
Paratype, Io.4632, female carapace	0·72	0·40	0·30
„ Io.4633, male left valve	0·95	0·49	
„ Io.4634, male right valve	0·95	0·47	

Remarks. C. levigata bears a close similarity to *Curfsina anorchidea* (Van Veen 1936), described from the Maastrichtian of Holland, but unlike that species does not taper so strongly towards the posterior end. In addition, the strongly serrated postero-ventral margin of *C. anorchidea* is a further distinguishing character. Apart from these features, however, the 2 species are very close morphologically, and both possess the same marginal swelling in front of the anterior socket of the left valve (text-figs. 30A, F). *Curfsina maior* (Van Veen 1936), selected as type species by Deroo (1966), has this same hinge development, but not so well developed as in *C. levigata*. This type of hinge is similar to the hinge present in *Idiocythere* Triebel, although the anterior swelling is not as pronounced, and does not take the form of an anterior tooth. The function is probably the same, however.

Genus CYTHEREIS Jones 1849
Cythereis brevicosta sp. nov.

Plate 20, figs. 9, 10; text-figs. 31A–D

1917   *Cythereis ornatissima* Reuss var. *nuda* Jones and Hinde; Chapman, p. 55, pl. 13, fig. 5; pl. 14, fig. 11.

Diagnosis. Cythereis with short median ridge not connected to sub-central tubercle. Intercostal regions smooth. Posterior end triangular, strongly tapered. Eye tubercle separated from dorsal ridge by smooth U-shaped area.

Holotype. Io.4605, right valve, Toolonga Calcilutite, Sample 4 (Santonian).

Paratypes. Io.4606–4607, 1 right and 1 left valve, Sample 4.

Description. Carapace large with high, well-rounded anterior margin. Eye tubercle prominently developed at anterior cardinal angle, which in left valve projects above right valve. Dorsal and ventral margins slope towards triangular posterior end.

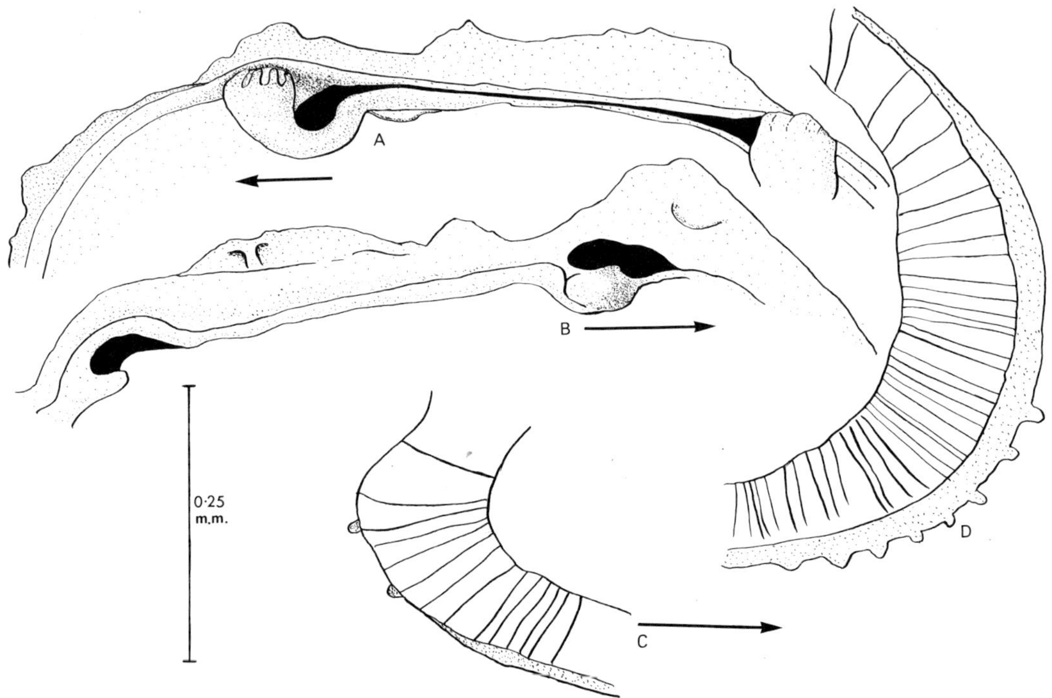

TEXT-FIGS. 31A–D. *Cythereis brevicosta* sp. nov. A, internal lateral view, right valve hinge, holotype Io.4605; B–D, internal lateral views, left valve hinge; posterior and anterior duplicature and radial pore canals, paratype Io.4607.

Postero-ventral and anterior margins coarsely dentate. Prominent, almost centrally situated tubercle separate from short median ridge, which may be broken down into discrete nodes. Low anterior marginal ridge continuous ventrally as corrugated ventral ridge. Dorsal ridge also has corrugated outline but separated from eye tubercle by smooth U-shaped area. Shell surface smooth.

Internally, hinge very strongly developed with anterior and posterior crenulations barely discernible. Although paramphidont, hinge (text-figs. 31A, B) approaches holamphidont condition. In holotype, small plate-like projection protrudes from anterior end of median element in right valve (text-fig. 31A).

Duplicature broad, with numerous, fine, radial pore canals; 36–40 anterior canals and 14–18 posterior canals observed (text-figs. 31C, D).

Muscle scars typical of genus but not clearly observed.

Dimensions (in mm)

	Length	Height
Holotype, Io.4605, right valve	1·00	0·56
Paratype, Io.4606, right valve	0·98	0·54
,, Io.4607, left valve	1·00	0·62
Chapman's material from Gingin:		
CPC.7135, right valve	0·83	0·44
CPC.7141, left valve	0·85	0·52

Remarks. The 2 specimens from the Gingin Chalk identified as *Cythereis ornatissima* var. *nuda* Jones and Hinde by Chapman (1917) are certainly not conspecific with the lectotype of *Cythereis nuda* Jones and Hinde (In.51685) in the British Museum Collections. They are, however, considered to be conspecific with the present material although somewhat smaller in size. Chapman's material tends to be more rectangular in shape, but this is probably related to the size difference.

C. brevicosta differs markedly from *C. nuda* in that the left valve has a distinct antero-dorsal projection at the anterior cardinal angle whilst the dorsal and ventral ridges are more positively developed. The postero-dorsal slope in *C. nuda* is also more noticeably concave than in *C. brevicosta*.

C. blanda Kaye (1963b), although similar because of its smooth intercostal surface, differs by not having such a well-developed antero-dorsal projection of the left valve, a more positive median ridge and a dorsal ridge which is in virtual continuity with the eye tubercle.

Genus HERMANITES s.l. Puri 1955

Hermanites sagitta sp. nov.

Plate 24, figs. 2–6; Plate 25, figs. 4, 5; text-figs. 32A–C

Diagnosis. Dimorphic species of *Hermanites* with broad, arrow-head outline in dorsal and ventral views. Hinge with all elements smooth. Inner margin not coincident with line of concrescence; narrow vestibule present along entire length of duplicature.

Holotype. Io.4635, female right valve, Korojon Calcarenite, Sample 2 (Campanian).

Paratypes. Io.4636–4642. 6 females and 1 male specimen, all Campanian: 2 females from the Korojon Calcarenite, Sample 1; 2 females and 1 male from Sample 2; and 2 females from the Toolonga Calcilutite, Sample 3.

Other material. 5 female specimens from Sample 2, and 3 specimens from the Toolonga Calcilutite, Sample 4 (Santonian).

Description. Carapace quadrate in female, more elongate in male. Eye tubercle prominent. Dorsal and ventro-lateral ridges distinct, projecting away from carapace at their posterior terminations. Carapace very broad posteriorly and ventrally flattened, tapering anteriorly to produce, in dorsal and ventral views, arrow-head outline. Sub-central swelling distinct. Shell surface coarsely reticulate. Anterior margin broadly rounded and dentate, especially antero-ventrally. Posterior end truncated, with deeply concave postero-dorsal slope and convex, dentate, postero-ventral slope. Left valve slightly larger than right.

Internally, hinge holamphidont, with all elements smooth.

Muscle scars consist of 4 oval adductor scars (imperfectly seen) and V- or heart-shaped antero-dorsal frontal scar (text-fig. 32A).

Inner margin does not coincide with line of concrescence, very narrow vestibule consequently developed along entire length of duplicature. Duplicature increases in width anteriorly and postero-ventrally. Radial pore canals long and simple, with only about 6 posteriorly and approximately 30 anteriorly. Selvage prominent in

E

both valves, subperipheral in left valve but set well back from valve margin in right valve.

Dimensions (in mm)

	Length	Height	Width
Holotype, Io.4635, female right valve	0·64	0·37	
Paratype, Io.4636, female carapace	0·73	0·43	0·45
„ Io.4637, female right valve	0·71	0·39	
„ Io.4638, female left valve	0·64	0·38	
„ Io.4639, female right valve	0·68	0·42	
„ Io.4640, female left valve	0·64	0·38	
„ Io.4641, male right valve	0·74	0·41	

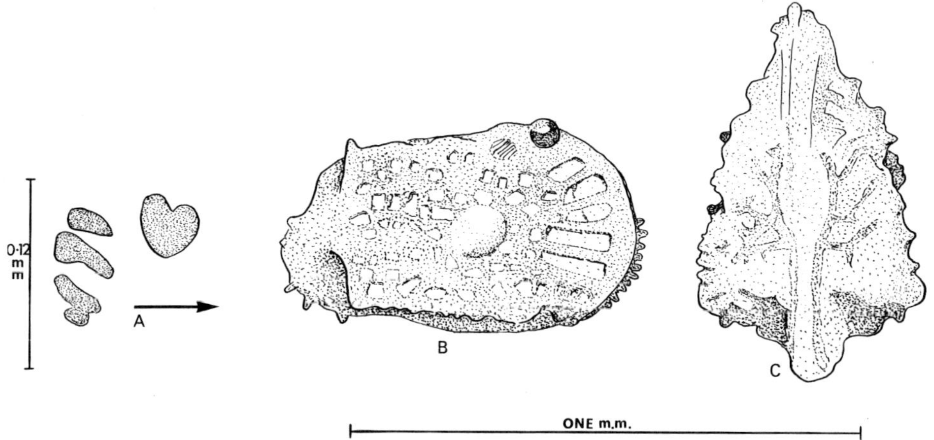

TEXT-FIGS. 32A–C. *Hermanites sagitta* sp. nov. A, muscle scars, female right valve, paratype Io.4637; B, male right valve, paratype Io.4641; C, ventral view, female carapace, paratype Io.4636.

Remarks. H. sagitta differs from other species of the genus in its distinct arrowhead shape in dorsal or ventral view. The absence of any dentition to the median hinge bar/groove and the presence of a narrow vestibule along the entire length of the duplicature contrasts this species from the generic diagnosis, which states that the median hinge bar/groove is dentate/loculate, and that a vestibule is not developed. *H. sagitta* is considered to be an early form of this essentially Tertiary to Recent genus.

Hazel (1967*b*) placed *Hermanites* in the Hemicytheridae, subfamily Thaerocytherinae. Apart from appendage characteristics, a feature of the Hemicytheridae is the presence of a divided frontal scar and in some instances division of 1 or more adductor scars, although this latter feature is not generally present in the Thaerocytherinae, a subfamily grouping of quadrate genera. Unfortunately the frontal scar of *Hermanites reticulata* (Puri), the type species, was not described by Puri (1953) and many species considered to belong to *Hermanites* possess a V-shaped or crescentic frontal scar, whilst others possess a divided frontal scar. Ruggieri (1962) gave an emended diagnosis for the genus in which he stated that there are

2 unequal scars in front of the adductor scars. The confusion which now exists will not be resolved until new material of the type species can be collected and examined. Then it may be necessary to split the genus into those possessing a V-shaped frontal scar and those having 2 oval frontal scars. Which type will be retained as *Hermanites* is conjectural.

As *H. sagitta* possesses a clear V-shaped frontal scar it is placed here in the Trachyleberideinae and is regarded as belonging to *Hermanites* s.l.

Genus KARSTENEIS Pokorny 1963

Subgenus KARSTENEIS Pokorny 1963

Remarks. Pokorny (1963) erected 2 subgenera for *Karsteneis*: *Karsteneis* (ventral and median ridges lacking, sub-central node absent or vestigial), and *Prosteneis* (3 distinct longitudinal ridges on the lateral surface, central node well developed).

Karsteneis (*K.*) *aspericava* sp. nov., although lacking a median ridge and possessing a very much reduced subcentral node, has rather prominent remnants of a dorsal and a ventro-lateral ridge. To this extent this form falls somewhat between the 2 subgenera of Pokorny. *K. aspericava* is, however, considered to be closer to *Karsteneis* than to *Prosteneis*.

Karsteneis (*Karsteneis*) *aspericava* sp. nov.

Plate 9, figs. 1–4; Plate 10, figs. 1–4; Plate 15, fig. 4; text-figs. 33A–C, 34A–C

Diagnosis. Karsteneis with coarsely pitted carapace. Remnants of dorsal and ventro-lateral ridges well developed in postero-dorsal and postero-ventral regions.

Holotype. Io.4481, carapace from the Toolonga Calcilutite, Sample 3 (Campanian).

Paratypes. Io.4482–4495. 14 specimens; 8 from Sample 3; 3 from the Korojon Calcarenite, Sample 1 (Campanian), and 3 from the Toolonga Calcilutite, Sample 4 (Santonian).

Other material. 14 specimens from Sample 1; 5 from Sample 2; 4 from Sample 3, and 3 from Sample 4.

Description. Carapace quadrate in female dimorph, more elongate in male. Juvenile instars more oval in outline, with distinct posterior taper. In adults anterior margin rounded; posterior margin triangular with concave postero-dorsal and convex postero-ventral slope. Anterior and posterior margins often dentate and in some instances even spinose (text-fig. 34C). Cardinal angles prominent and distinct eye node situated beneath anterior angle. Sub-central tubercle weakly developed, and should perhaps be termed sub-central swelling. Shallow sulcus behind swelling. Ventro-lateral area of carapace developed into lateral ridge, particularly prominent at its posterior termination; also true for impersistently developed dorsal ridge.

Shell surface very coarsely pitted towards centre of carapace and on ventral surface. Size of pits of lateral surface decreases outwards towards terminal margins. Large oval normal pore canal openings with sieve plate made up of biserial row of elongate pores separated by central bar (Pl. 15, fig. 4). Left valve only slightly larger than right.

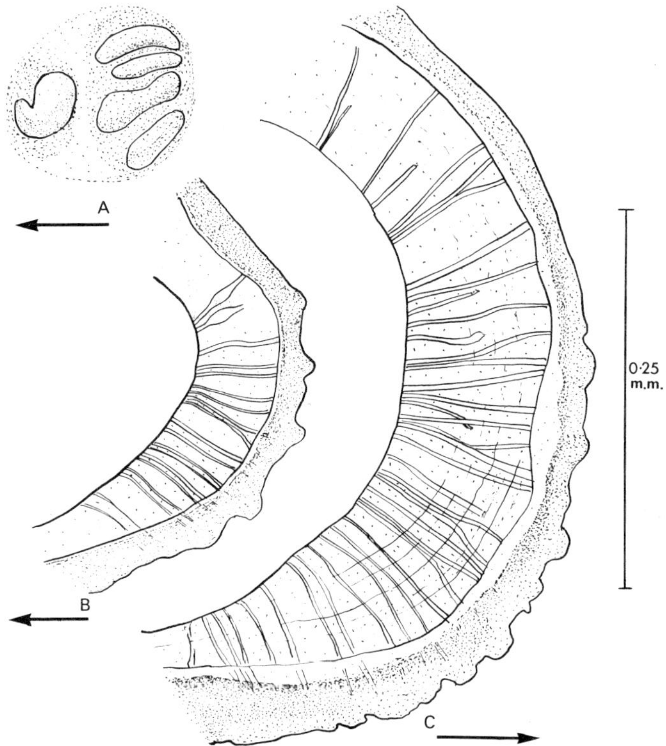

TEXT-FIGS. 33A–C. *Karsteneis* (*Karsteneis*) *aspericava* sp. nov.
A, muscle scars, female right valve, paratype Io.4495; B, posterior
margin with duplicature and radial pore canals, female right
valve, paratype Io.4490; C, anterior margin showing dentate
outline, and duplicature with radial pore canals, female left valve,
paratype Io.4491.

Hinge holamphidont; median bar and tooth of left valve smooth; in right valve posterior, blade-like tooth and anterior knob-like tooth both very weakly scored, but do not approach paramphidont condition. Pokorny (1963) stated that both terminal teeth notched, but did not explain further or illustrate this feature. Hinge as developed here probably closer to holamphidont condition than in Pokorny's species.

Muscle scars (text-fig. 33A) within sub-central node or swelling: 4 rather elongate adductor scars separated from V-shaped frontal scar by low ridge.

Inner margin and line of concrescence coincide to produce rather broad duplicature, across which pass 30–39 anterior and 17–19 posterior radial pore canals (text-figs. 33B, C), long, often grouped in pairs, and of sinuous habit. Some apparently branch but actually case of 2 closely positioned canals diverging. Some canals open on to lateral surface and often termed false pore canals; they start from same position, however, and bristles they contained almost certainly had same function.

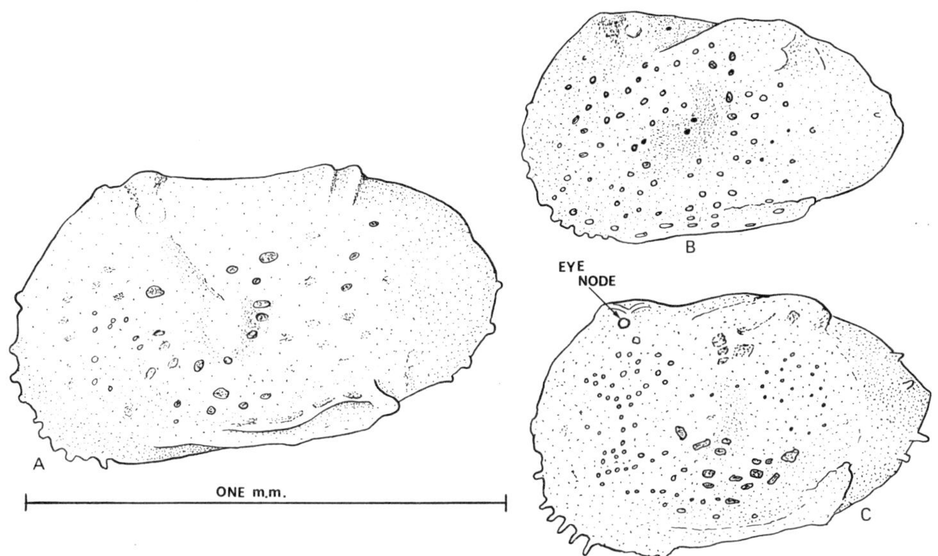

TEXT-FIGS. 34A–C. *Karsteneis (Karsteneis) aspericava* sp. nov.   A, left side, male carapace, paratype Io.4487; B, left side, female carapace showing slight variations in carapace outline and degree of marginal spinosity when compared with holotype; paratype Io.4493; C, left side of female carapace to show marginal spines, holotype Io.4481.

Dimensions (in mm)

	Length	*Height*	*Width*
Holotype, Io.4481, female carapace	0·83	0·52	0·43
Paratype, Io.4482, female right valve	0·79	0·49	
,, Io.4483, female left valve	0·80	0·51	
,, Io.4484, female right valve	0·79	0·46	
,, Io.4485, juvenile right valve	0·63	0·42	
,, Io.4486, juvenile left valve	0·66	0·42	
,, Io.4487, male carapace	0·95	0·56	0·47
,, Io.4488, male right valve	0·88	0·49	
,, Io.4489, female carapace	0·88	0·55	0·44
,, Io.4490, female right valve	0·82	0·49	
,, Io.4491, female left valve	0·79 (broken)	0·52	
,, Io.4492, female left valve	0·77	0·47	
,, Io.4493, female carapace	0·77	0·48	0·40
,, Io.4494, male right valve	0·83	0·43	
,, Io.4495, female right valve	0·81	0·49	

Remarks. Cythereis ornatissima Reuss var. *nuda* Jones and Hinde as figured and described by Chapman (1917) (i.e. the *Cythereis brevicosta* of this paper) is very close to the male dimorph of *K.* (*K.*) *aspericava*, and great care must be taken in separating these 2 forms. *K. aspericava* differs by having a coarsely pitted lateral surface, no median ridge, and less well-developed eye and sub-central nodes. The dorsal and ventro-lateral ridges present in Chapman's *C. ornatissima nuda* material are only poorly developed in *K. aspericava*.

Neither of Pokorny's species, *K.* (*Karsteneis*) *karsteni* and *K.* (*Prosteneis*) *nodifera*, possess the coarse pitting of *K. aspericava*.

Genus LIMBURGINA Deroo 1966

Remarks. Externally *Limburgina* resembles the genus *Cythereis* but internally differs in the possession of a hemiamphidont hinge. In the type species, *Limburgina ornata* (Bosquet), the posterior hinge element on the right valve is only weakly subdivided into a tri-lobed tooth; the hinge is close to being holamphidont. The V-shaped frontal scar of the trachyleberids is present.

Limburgina formosa sp. nov.

Plate 23, figs. 1–8; text-figs. 35A–F

Diagnosis. Large ornate, spinose species of *Limburgina* with coarse reticulate ornamentation. Short lateral spines project from reticulae into surface pits. All margins of carapace spinose; spines also present at nexus of reticulae.

Holotype. Io.4643, male left valve, Toolonga Calcilutite, Sample 3 (Campanian).

Paratypes. Io.4644–4650. 2 female valves, Sample 3; 3 male and 1 female valve, and 1 juvenile carapace, Korojon Calcarenite, Sample 2 (Campanian).

Other material. 1 specimen, Korojon Calcarenite, Sample 1 (Campanian) and 3 specimens, Sample 2.

Description. Carapace large, rectangular, with high, broadly rounded anterior margin, almost parallel dorsal and ventral margins and broadly triangular posterior end. Sexual dimorphism indicated by presence of more elongate males. Left valve larger than right. Anterior and postero-ventral margins possess long marginal

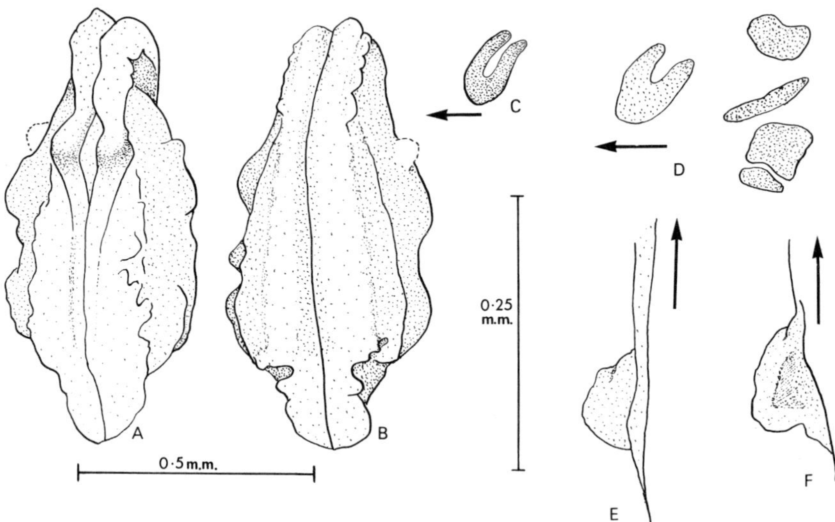

TEXT-FIGS. 35A–F. *Limburgina formosa* sp. nov. A, B, dorsal and ventral views, juvenile carapace, paratype Io.4647; C, U-shaped frontal muscle scar, female right valve, paratype Io.4649; D–F, muscle scars, anterior hinge tooth and posterior hinge tooth, male right valve, paratype Io.4646.

spines. Spines along dorsal and ventral margins shorter but appear to have tips broken off; true also for spines at nexus of coarse reticulate surface ornament. All spines appear to be hollow but probably not so, terminal pits being produced when spine tips broken off.

Within reticulate ornament, small lateral spines grow out from reticulae into surface pits (Pl. 23, figs. 2, 7), and small normal pore canals open along ridges. This surface ornamentation superimposed upon generic features of large antero-central swelling and 3 lateral ridges. Ridges not prominently developed and occur in dorsal, median, and ventro-lateral position.

Large eye node just below anterior cardinal angle (Pl. 23, fig. 9) and opens internally just in front of anterior hinge element (Pl. 23, fig. 3).

Internally, hinge well developed and amphidont, having weakly notched anterior and posterior teeth (text-figs. 35E, F) in right valve. Left valve has weakly stepped socket at posterior end only (Pl. 23, fig. 4), anterior socket being quite smooth (Pl. 23, fig. 6). Median bar in left valve also smooth (Pl. 23, fig. 5). Hinge may be described as paramphidont, but clearly not very far removed from holamphidont type.

Inner margin and line of concrescence coincide and moderately broad duplicature present. Radial pore canals long and straight, approximately 40 anteriorly and 10 or 11 posteriorly.

Muscle scars difficult to determine, but resolved as vertical row of 4 adductor scars with oblique U-shaped frontal scar (text-figs. 35C, D) folded back upon itself and which at first glance often appears to be simple oval scar.

Dimensions (in mm)

	Length	Height	Width
Holotype, Io.4643, male left valve	1·18	0·69	
Paratype, Io.4644, female left valve	1·08	0·69	
,, Io.4645, female left valve	1·10	0·69	
,, Io.4646, male right valve	1·15	0·62	
,, Io.4647, juvenile carapace	0·86	0·49	0·40
,, Io.4648, male right valve	1·15	0·62	
,, Io.4650, male right valve	1·10	0·60	

Remarks. The ostracod described by Chapman (1917) as *Cythereis ornatissima* Reuss, is closely similar to *L. formosa* but differs in the right valve by having a rather narrow lateral outline and a more distinctly tapered posterior end, with the line of greatest length situated slightly below the mid-point instead of passing through it. The anterior margin also has an antero-ventral projection whilst the fragmented dorsal ridge has a short vertical ridge connecting it to the antero-central swelling, features absent from *L. formosa*. The 2 forms could be related, however, and Chapman's species occupies an ancestral position within the Santonian Gingin Chalk.

The broadly triangular posterior end of *L. formosa*, with its simple, straight spines projecting from it, contrasts with the rather heavy, coarsely spinose postero-ventral margins present in the majority of European species of this genus.

Genus OERTLIELLA Pokorny 1964
Oertliella exquisita sp. nov.

Plate 24, fig. 1; Plate 25, figs. 1–3, 6, 7; text-figs. 36A–G

Diagnosis. Oertliella with neat reticulate ornament. Ventro-lateral ridge well developed, curving upwards posteriorly.

Holotype. Io.4651, female left valve, Toolonga Calcilutite, Sample 3 (Campanian).

Paratypes. Io.4652–4657. 4 female and 2 male specimens, Sample 3.

Other material. 9 specimens from the Korojon Calcarenite: 6 from Sample 1 and 3 from Sample 2 (Campanian); together with 10 specimens from Sample 3.

Description. Carapace rectangular, with slightly more elongate male dimorphs. Anterior margin broadly rounded and spinose. Posterior end broadly triangular with concave postero-dorsal slope (especially in right valve) and convex, coarsely spinose postero-ventral slope. Ventral margin very slightly convex with unusual, flattened incurvature (text-fig. 36A) in right valve. Dorsal margin straight, sloping to posterior end.

Prominent eye tubercle opens internally in front of anterior hinge element as prominent ocular pit; externally, tubercle supported on projecting pillar at anterior cardinal angle.

Sub-central tubercle clearly visible but not particularly well developed. Dorsal ridge rather fragmentary, terminating at its posterior end in dorsally projected spine.

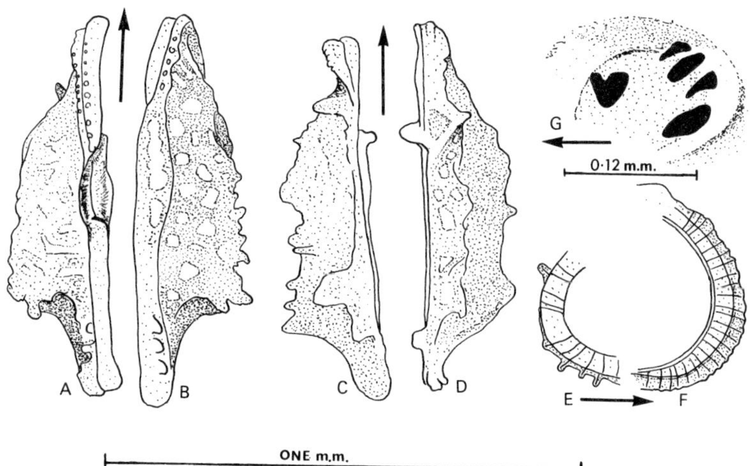

TEXT-FIGS. 36A–G. *Oertliella exquisita* sp. nov. A, D, ventral view, showing unusual flattened plate-like flange in antero-median position and dorsal view to show hinge teeth, female right valve, paratype Io.4652; B, C, ventral and dorsal views, female left valve, holotype Io.4651; E, F, posterior and anterior duplicatures showing radial pore canals and anterior vestibule, female left valve, paratype Io.4656; G, muscle scars situated in muscle scar pit, female right valve, paratype Io.4655.

Ventro-lateral ridge well developed, spinose, curving noticeably upwards posteriorly. In dorsal and ventral views this ridge projects laterally from carapace (text-figs. 36A–D). Entire shell surface covered by neat reticulate ornament.

Normal pore canals open within pits as circular openings within slightly upraised 'mound'. Each canal surrounded by numerous smaller pores (Pl. 25, fig. 6).

Internally, hinge holamphidont, although some very weak crenulation of posterior tooth (right valve) observeable in favourable light.

Muscle scars (text-fig. 36G) within deep pit; V-shaped frontal scar well developed.

Anteriorly, duplicature not very broad, and narrow vestibule (text-fig. 36F) is present. 24 anterior and 11 posterior, straight, radial pore canals (text-figs. 36E, F).

Selvage peripheral in left valve but bordered by broad flange in right valve, well developed (Pl. 25, fig. 3) around anterior margin and along ventral margin, especially in posterior half. Antero-medially, flange flattened at right angles to normal flange projection (text-fig. 36A).

Dimensions (in mm)

	Length	*Height*
Holotype, Io.4651, female left valve	0·77	0·44
Paratype, Io.4652, female right valve	0·77	0·42
,, Io.4653, male left valve	0·84	0·43
,, Io.4654, male right valve	0·82	0·43
,, Io.4655, female right valve	0·78	0·42
,, Io.4656, female left valve	0·77	0·45

Remarks. O. exquisita is closely similar to *Oertliella reticulata* (Kafka 1886, see Pokorny 1964), the type species, but may be distinguished by the straight ventro-lateral ridge of *O. reticulata* and the upwardly curving ridge of *O. exquisita*.

Oertliella binkhorsti (Van Veen 1936, see Herrig 1966) is a more quadrate species with a straight ventro-lateral ridge and a much coarser reticulate ornament.

Genus SCEPTICOCYTHEREIS nov.

Type species. Scepticocythereis ornata sp. nov.

Diagnosis. Rectangular to quadrate genus of *Cythereis*-type appearance. Anterior end rounded, posterior end triangular; margins thickened, spinose. Shell surface reticulate. Sub-central swelling not distinct; lateral ridges poorly developed, median ridge restricted to posterior half of carapace, dorsal and ventral ridges discontinuous. Eye tubercle either strongly or poorly developed. Hinge hemimerodont in juvenile instars; amphidont in adult instars. Duplicature very narrow; line of concrescence and inner margin coinciding. Radial pore canals short, straight, few. Muscle scars subvertical row of 4 unequal adductor scars, an antero-median oval or slightly crescentic frontal scar and rounded to angular antero-ventral mandibular scar.

Remarks. Scepticocythereis is basically a *Cythereis*-type genus virtually lacking the 3 lateral ridges. The characteristic antero-dorsal elevation of the left valve of *Cythereis* is also absent, as also is a well-developed anterior marginal ridge. The muscle scar pattern, with an elongate frontal scar, contrasts *Scepticocythereis* with the other members of the Trachyleberideinae, and it is only questionably included in this subfamily. It is possible, however, that the short arm of the frontal scar required to convert the present form into a V-shaped scar is simply not preserved in the material available.

Murrayina Puri (1953) from the Miocene of Florida is a closely similar genus having a reticulate shell surface and lacking ridges. Although the lateral ridges of *Scepticocythereis* are very poorly developed, they are, nevertheless, still discernible, and the broad triangular posterior end serves to increase the differences between this genus and the bluntly rounded *Murrayina*.

Karsteneis Pokorny (1963) is another form erected for those trachyleberids which have no or very poorly developed lateral ridges. It is, however, a very different genus which is unlikely to be confused with *Scepticocythereis*.

Scepticocythereis ornata sp. nov.

Plate 26, figs. 1–8; Plate 27, figs. 11, 12; text-figs. 37A–F

1917   *Cythereis ornatissima* Reuss var. *reticulata* Jones and Hinde; Chapman, p. 55, pl. 14, fig. 12.

Diagnosis. Scepticocythereis with neat reticulate ornament in which small spines extend into surface pits. Median ridge represented by small spinose development in posterior half of carapace; dorsal ridge a series of isolated, short spines; ventro-lateral ridge more positive but tending to consist of series of discontinuous spines.

Holotype. Io.4658, male left valve, Toolonga Calcilutite, Sample 4 (Santonian).

Paratypes. Io.4659–4668, 3 females, 3 males, and 4 juveniles, Sample 4.

Other material. 12 specimens, Sample 4.

Description. Carapace rectangular in male dimorph, slightly more quadrate in female. Anterior end high with broadly rounded, spinose margin; posterior end narrower and triangular with spines along postero-ventral margin.

Juvenile instars similar to adults except for much stronger posterior taper and tendency to develop distinct postero-dorsal projection (Pl. 26, fig. 4).

In adults, dorsal margin straight, sloping to posterior; ventral margin very gently convex with shallow antero-median incurvature. Left valve slightly larger than right, with noticeable overlap along ventral margin.

Shell surface neatly reticulate, with short spines extending into pits from reticulae (Pl. 26, fig. 3; Pl. 27, fig. 12); this development less noticeable in juvenile instars.

Eye tubercle somewhat variable in degree of development (text-figs. 37A, B) at or slightly below anterior cardinal angle.

Sub-central tubercle present with noticeable carapace constriction behind. Of 3 lateral ridges, median ridge represented solely by small spinose development in posterior half of carapace, dorsal ridge by somewhat irregular row of small discrete spines, and ventro-lateral ridge by more positive development of spines (Pl. 27, fig. 12) curving slightly upwards posteriorly.

Internally, hinge hemimerodont in juvenile instars (Pl. 26, figs. 6–8) but develops into amphidont condition in adult instars (text-figs. 37A–C), with some crenulation of terminal elements remaining. In 1 female valve (text-figs. 37D, E), although the antero-median socket well formed, terminal teeth still not fully developed amphidont type, as present in large males.

Muscle scars within muscle scar pit with usual 4 oval but unequal adductor scars and somewhat angular antero-ventral mandibular scar. Frontal scar seen in number

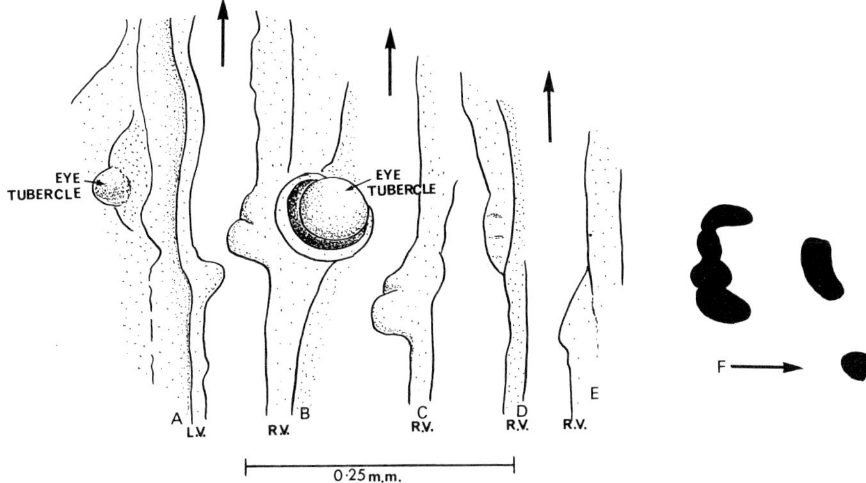

TEXT-FIGS. 37A–F. *Scepticocythereis ornata* gen. et sp. nov. A, dorsal view, showing antero-median hinge tooth and small eye tubercle, male left valve, holotype Io.4658; B, C, F, dorsal view, showing anterior and posterior hinge teeth and large antero-dorsal eye tubercle, and muscle scars, male right valve, paratype Io.4660; D, E, dorsal view, female right valve, showing anterior and posterior hinge teeth, paratype Io.4661.

of individuals always an elongate scar, sometimes slightly crescentic (text-fig. 37F) but never V-shaped.

Inner margin and line of concrescence coincide to produce very narrow duplicature. No vestibule. Radial pore canals short, straight, probably no more than 15 anteriorly; precise number not determined.

Dimensions (in mm)

	Length	Height	Width
Holotype, Io.4658, male left valve	0·93	0·49	
Paratype, Io.4659, male carapace	0·93	0·49	0·46
„ Io.4660, male right valve	0·93	0·49	
„ Io.4661, female right valve	0·82	0·45	
„ Io.4662, juvenile right valve	0·66	0·37	
„ Io.4663, juvenile left valve	0·68	0·41	
„ Io.4664, female right valve	0·88	0·48	
„ Io.4665, female carapace	0·83	0·49	0·43
„ Io.4666, juvenile left valve	0·67	0·40	
„ Io.4667, male left valve	0·93	0·49	
„ Io.4668, juvenile right valve	0·69	0·37	

Remarks. The figured specimen of Chapman's *Cythereis ornatissima reticulata* (CPC.7142) from the Gingin Chalk is not conspecific with Jones and Hinde's species of that name, which was described from the Gault of England and the Chalk of both England and Northern Ireland. Chapman's species is conspecific with *S. ornata* and the figured specimen is a male carapace.

S. ornata appears to be restricted to sediments of Santonian age only.

Genus TOOLONGELLA nov.

Type species. Toolongella mimica sp. nov.

Diagnosis. Trachyleberid ostracod genus with quadrate carapace ornamented by 3 parallel postero-lateral ridges; distinct sub-central tubercle; anterior marginal ridge which may extend back along ventro-lateral margin. Normal pore canal apertures large, few. Anterior and posterior margins often dentate. Hinge holamphidont. Muscle scars with V-shaped frontal scar. Duplicature without vestibule; 25–34 anterior and 15–20 posterior, straight, radial pore canals. Left valve slightly larger than right.

Remarks. Toolongella is closely similar to *Sergipella* Krömmelbein (1967), although it differs in the possession of an increased number of anterior radial pore canals and in the fact that the lowermost lateral ridge does not posteriorly turn down to fuse with a ventral ridge. The 3 lateral ridges of *Toolongella* are quite separate. The close similarity of these 2 genera suggests a possible phylogenetic link, and the Aptian/Albian *Sergipella* from Brazil might well be ancestral to the Santonian/Campanian *Toolongella.*

Toolongella mimica sp. nov.

Plate 19, figs. 5, 8–10; Plate 20, figs. 1–4; text-figs. 38A–C, 39A–C

Diagnosis. Toolongella with 3 short, postero-lateral ridges; large sub-central tubercle with vertical swelling in front. Anterior marginal ridge, when present, terminates in low swelling at anterior cardinal angle. Ventral surface flattened with irregular ventro-lateral margins diverging towards posterior end, then finally narrowing. Carapace strongly convex in dorsal view. Marginal denticles often present. Shell surface with very fine striae producing reticulate pattern. Small, flattened projection in postero-ventral region. Greatest length of carapace situated below mid-point; postero-dorsal slope long, steeply angled.

Holotype. Io.4593, right valve, Toolonga Calcilutite, Sample 3 (Campanian).

Paratypes. Io.4594–4601, Io.4604; from the Toolonga Calcilutite: 5 specimens from Sample 3 (Campanian) and 1 from Sample 4 (Santonian); from the Korojon Calcarenite: 1 specimen from Sample 1 and 2 from Sample 2, both Campanian.

Other material. 1 specimen, Sample 1; 19 specimens, Sample 2, and 1 specimen, Sample 3.

Description. Carapace quadrate, projected dorsally at anterior cardinal angle. Anterior margin broadly curved, dentate; posterior end triangular with long, steeply sloping postero-dorsal slope and short, convex, dentate, postero-ventral slope. Greatest length of carapace passes below mid-point. Carapace rather convex in dorsal view with left valve antero-dorsally overlapping right. Ventrally, left valve overlaps right along ventral margin.

Shell surface ornamented by 3 short, postero-lateral ridges, uppermost of which distinctly swollen in dorsal view. In front of well-developed sub-central tubercle, a short, rather swollen vertical ridge lies inside curvature of anterior marginal swelling.

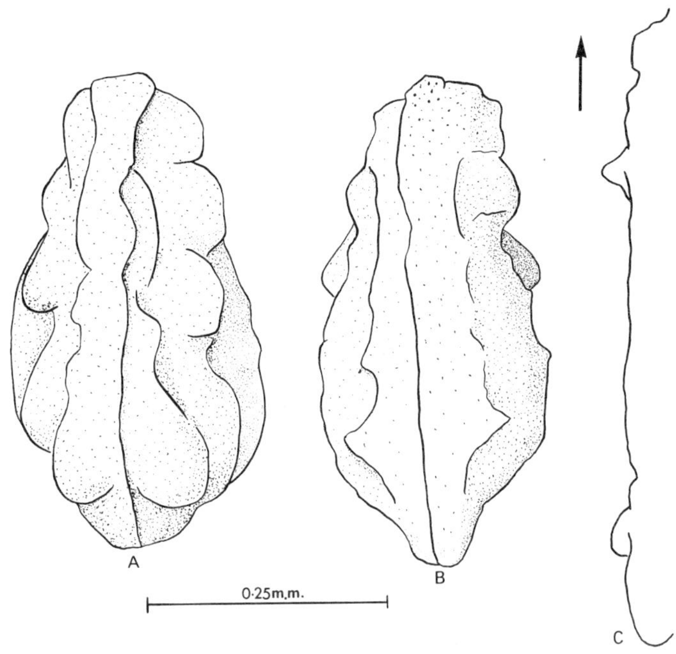

TEXT-FIGS. 38A–C. *Toolongella mimica* gen. et sp. nov. A, B, dorsal and ventral views, complete carapace, paratype Io.4594; C, dorsal view, right valve hinge; paratype Io.4604.

Anterior marginal ridge not always present but when observed terminates at antero-dorsal angle in low swelling. In 1 specimen (Pl. 19, fig. 8) ridges broken down into series of discrete tubercles and rather unusual flattened postero-ventral projection of species particularly well developed; this specimen Santonian not Campanian as others. Shell surface additionally ornamented by very fine reticulation, striae which produce it being only clearly seen in well-preserved specimens. Normal pore canal apertures large, circular, relatively few.

Internally, hinge holamphidont (Pl. 19, fig. 5; text-fig. 38c), all elements being smooth. Anterior tooth of right valve cone-shaped; posterior tooth very much broader. Duplicature of moderate width, without vestibule. 25–34 straight anterior radial pore canals and approximately 17 posterior canals (text-figs. 39B, C). Variation in number anteriorly may be partly due to state of preservation or may represent true variation.

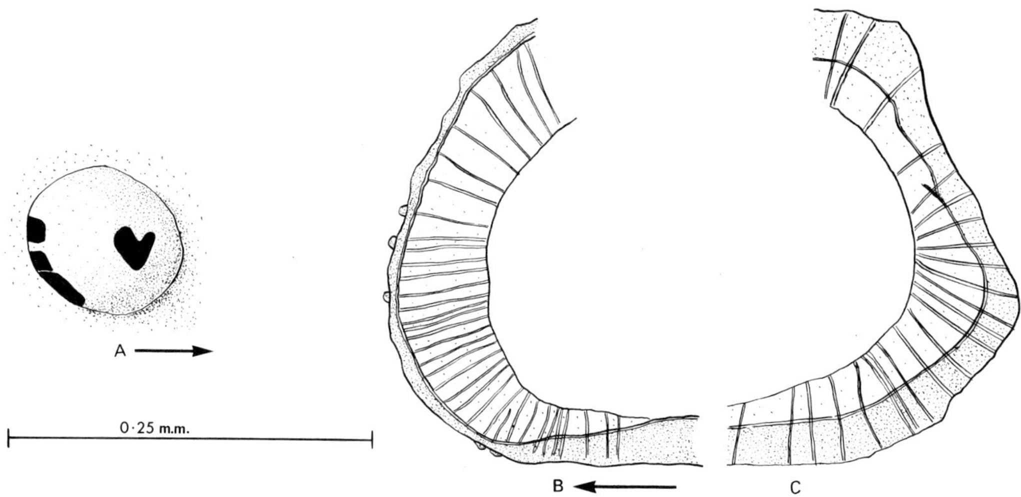

TEXT-FIGS. 39A–C. *Toolongella mimica* gen. et sp. nov.   A, muscle scars as viewed from the exterior, right valve, paratype Io.4596; B, C, anterior and posterior margins to show duplicature and radial pore canals, right valve, holotype Io.4593.

Muscle scars inside sub-central tubercle pit and only observed from exterior (text-fig. 39A). V-shaped frontal scar, however, well developed.

Dimensions (in mm)

	Length	Height	Width
Holotype, Io.4593, right valve	0·60	0·34	
Paratype, Io.4594, carapace	0·52	0·32	0·26
,, Io.4595, left valve	0·54	0·35	
,, Io.4596, right valve	0·59	0·36	
,, Io.4597, right valve	0·50	0·30	
,, Io.4598, left valve	0·54	0·34	
,, Io.4599, left valve	0·50	0·29	
,, Io.4600, right valve	0·59	0·35	
,, Io.4601, left valve	0·54	0·32	
,, Io.4604, right valve	0·66	0·37	

Remarks. T. mimica has some external similarity to *Mauritsina hieroglyphica* (Bosquet 1962, see Deroo 1962), but has a completely different muscle scar pattern. The presence in the Santonian of a form having the lateral ridges broken up into discrete tubercles is regarded as a variant at the present time. Whether the Santonian form will later be split off as a species in its own right is dependent on further material being found.

?*Toolongella* sp.

Plate 19, figs. 6, 7, 11; text-fig. 40

Remarks. 3 specimens of a distinct species having a close similarity to *Toolongella* have been obtained from the Toolonga Calcilutite, Sample 4 (Santonian). The main ornamentation consists of 3 lateral ridges which are more elongate than those in

T. mimica. The anterior marginal ridge is also continuous around the ventral, posterior, and dorsal margins. The dorsal part of this ridge turns down anteriorly to extend into the central part of the valve and terminates in a small node in front of and below the sub-central tubercle.

The extension of the anterior marginal ridge around the valve margins and the turning down into the anterior part of the valve of the dorsal ridge clearly distinguishes this species from *T. mimica.* A further distinguishing feature is the rather rounded posterior margin, which contrasts markedly with the triangular posterior end of

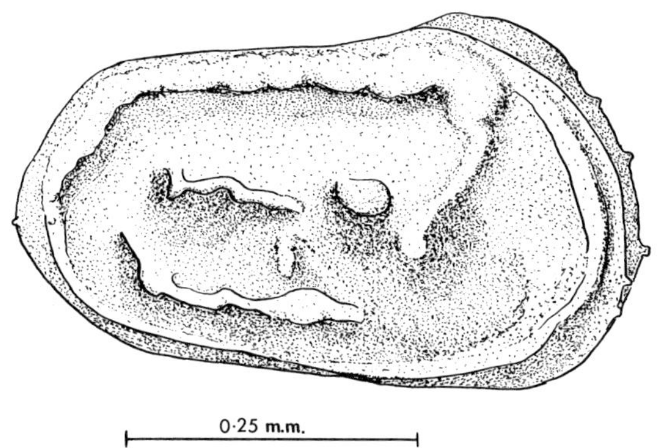

TEXT-FIG. 40. ?*Toolongella* sp., right side of complete
carapace, Io.4602.

T. mimica. The rather distinctive carapace ornamentation suggests that this species may represent a new genus. Until additional material becomes available this, together with the specific identification, must remain open.

Dimensions (in mm)

	Length	Height	Width
Io.4602, carapace	0·54	0·32	0·26
Io.4603, left valve	0·50	0·29	
Io.4609, right valve	0·50	0·29	

Genus TRACHYLEBERIDEA Bowen 1953 emend. Hazel 1965
Trachyleberidea sp.

Plate 22, fig. 6

Remarks. 3 specimens have been found in the Upper Gearle Siltstone, Sample 5 (Coniacian). The largest specimen (figured) is slightly damaged along the postero-dorsal slope and the remaining 2 specimens do not provide sufficient detail for specific description.

Dimensions. Io.4627, left valve, length 0·65 mm, height 0·34 mm.

Genus TRACHYLEBERIS Brady 1898

Remarks. The genus *Trachyleberis* Brady is generally regarded as having a geological range from Palaeocene to Recent. In Western Australia, however, 3 species of *Trachyleberis* have been identified: 2 from the Coniacian Upper Gearle Siltstone and 1 from the Campanian Toolonga Calcilutite and Korojon Calcarenite.

Trachyleberis sp., Type A

Plate 20, fig. 8; Plate 22, fig. 5

Remarks. A number of fragmented specimens occur in the Upper Gearle Siltstone, Sample 6 (Coniacian). The spines are terminally subdivided into a number of smaller spines and the eye tubercle is prominent.

Material. Io.4610, anterior fragment, left valve. Io.4611, posterior fragment, left valve.

Trachyleberis sp., Type B

Plate 22, fig. 4; Plate 23, fig. 9

Remarks. 1 complete left valve and a number of fragments, from the Upper Gearle Siltstone, Sample 5 (Coniacian). Although similar to Type A above, the spines in Type B are not so well developed and a prominent eye tubercle is lacking. A very weak reticulation can be seen on the smooth surface of the shell, between the spines.

Dimensions. Io.4612, left valve, length (broken), 0·77 mm, height 0·45 mm. Io.4613, right valve, length 0·83 mm, height 0·43 mm.

Trachyleberis anteplana sp. nov.

Plate 20, figs. 5–7; Plate 22, figs. 1–3; text-figs. 41A, B

1917   *Cythereis rudispinata* Chapman and Sherborn; Chapman, p. 56, pl. 14, fig. 15.

Diagnosis. Trachyleberis with long marginal spines, and broad spine-free area behind anterior margin. Eye tubercle prominent with short spine situated behind and projecting obliquely backwards. Shell surface often weakly reticulate; reticulae produced by very fine striae.

Holotype. Io.4614, female right valve, Toolonga Calcilutite, Sample 3 (Campanian).

Paratypes. Io.4615–4620, 6 specimens: 3 specimens from Sample 3, and 3 from the Korojon Calcarenite, Sample 2 (Campanian).

Description. Carapace has high and broadly rounded anterior end and tapers to narrow, triangular posterior end. More elongate forms considered to be male dimorphs. Dorsal margin straight, sloping obliquely backwards towards concave postero-dorsal slope. Anterior margin possesses double row of long spines, inner row of which continuous ventrally with row of spines which laterally passes up on to ventro-lateral flank of carapace. Postero-ventral margin also strongly spinose. Laterally, 2 main clusters of spines: firstly in region of central trachyleberid tubercle, and secondly in comparable position in posterior half of valve. Large spines along dorsal margin, but especially at postero-dorsal angle. Short, backwardly projected spine immediately above and behind prominent eye tubercle.

In Toolonga Calcilutite specimens, reticulate network of very fine striae seen on otherwise smooth areas of shell surface, between spines (Pl. 20, fig. 6).

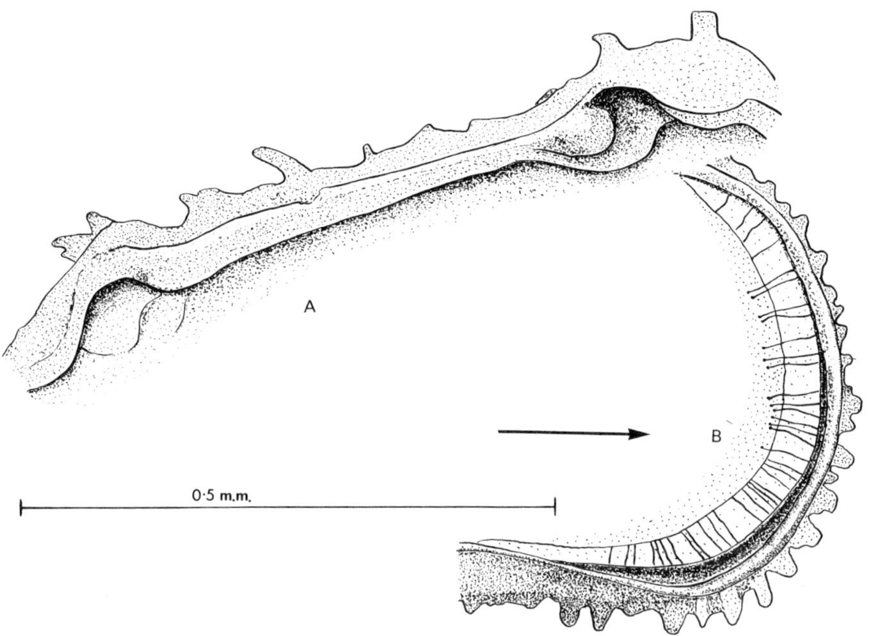

TEXT-FIGS. 41A, B. *Trachyleberis anteplana* sp. nov. A, left valve hinge, male, paratype Io.4620; B, anterior margin showing dentate outline, duplicature and radial pore canals; note inward extension of pore canals at mid-height beyond the inner margin, broken left valve, paratype Io.4616.

Broad spine-free area just behind inner row of anterior marginal spines; in absence of fine striae, area perfectly smooth and characterizes species.

Internally hinge holamphidont; all elements smooth (text-fig. 41A). Inner margin and line of concrescence coincide to produce relatively narrow duplicature. Approximately 38 straight or slightly sinuous anterior radial pore canals observed, canals being grouped in pairs above mid-height, but more closely arranged below mid-height. At mid-height, anterior radial pore canals seen to extend inwards beyond inner margin (text-fig. 41B). Posterior canals not observed.

Muscle scars (Pl. 22, fig. 1) have well-developed V-shaped frontal scar, together with elongate adductor scars.

F

Dimensions (in mm)

	Length	Height
Holotype, Io.4614, female right valve	0·93	0·54
Paratype, Io.4615, male right valve	1·00	0·51
„ Io.4617, female right valve	0·95	0·53
„ Io.4618, male left valve	1·20	0·61
„ Io.4619, female right valve	0·89	0·57
„ Io.4620, male left valve	1·03	0·60

Remarks. Although coated with matrix, Chapman's Gingin Chalk species, *Cythereis rudispinata* Chapman and Sherborn (Chapman 1917), is considered to be synonymous with *T. anteplana*. Chapman's species is not, however, synonymous with Chapman and Sherborn's (1893) *C. rudispinata* from the Gault of Folkestone.

The Coniacian *Trachyleberis* sp. (Type B) is close to *T. anteplana* but differs in having a less prominent eye tubercle and shorter marginal spines, which, as far as the inner row is concerned, are thickened at their bases to produce a marginal ridge. These morphological differences may be observed in the illustrations of the 2 species (Pl. 20, figs. 5, 6; Pl. 22, fig. 4).

The extension internally of some of the anterior radial pore canals beyond the line of concrescence is an unusual feature and indicates that the canals are here within the outer lamella rather than simply passing through the plane of fusion of the inner and outer lamellae.

Family uncertain
Genus HYSTRICOCYTHERE nov.

Type species. Hystricocythere imitata sp. nov.

Diagnosis. Small, ovoid genus, highest at anterior end. Slight mid-dorsal and mid-ventral constriction in lateral view. Carapace slender in dorsal view. Convex postero-dorsal projection. Shell surface reticulate, spinose. Hinge hemimerodont. Muscle scars with 4 oval adductor scars and oval antero-dorsal frontal scar. Inner margin and line of concrescence coincide. Radial pore canals few, long, straight. Left valve slightly larger than right.

Remarks. Hystricocythere is a monotypic genus, the type species of which is small and spinose. Externally there is some resemblance to the genus *Echinocythereis* Puri (1953) but the eye node and double frontal scar of the latter are not present, and there are also fewer radial pore canals and a different hinge.

The family assignment of this genus is not definitely known at present and it is preferred to retain it in an unclassified unit.

Hystricocythere imitata sp. nov.

Plate 24, fig. 7; Plate 27, figs. 1–10; text-figs. 42A, B

1917   *Cythere harrisiana* Jones var. *reticosa* Jones and Hinde; Chapman, p. 53, pl. 13, fig. 6.

Diagnosis. Hystricocythere with coarse reticulate ornament; reticulae bear haphazard arrangement of short, stubby spines at valve centre; spines arranged in rows towards valve margins. Approximately 14 anterior radial pore canals.

Holotype. Io.4669, right valve, Toolonga Calcilutite, Sample 4 (Santonian).

Paratypes. Io.4670–4678. 8 specimens from Sample 4; 1 specimen from the Toolonga Calcilutite, Sample 3 (Campanian).

Other material. 1 specimen from Sample 3, and 5 from Sample 4.

Description. Carapace small, ovoid, with medially concave dorsal and ventral margins. Internally, hinge margin of both valves perfectly straight. Anterior end high, broadly rounded, spinose. Carapace tapers towards narrowly rounded and

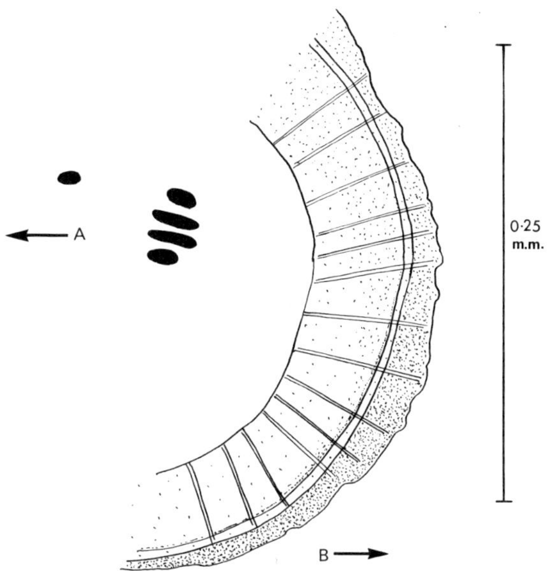

TEXT-FIGS. 42A, B. *Hystricocythere imitata* gen. et sp. nov. A, muscle scars, left valve, paratype Io.4673; B, anterior margin with duplicature and radial pore canals, left valve, paratype Io.4677.

spinose posterior end. Postero-dorsal region of each valve distinctly convex, projecting above dorsal margin in lateral view. Left valve slightly larger than right, with overlap at anterior cardinal angle and along ventral margin.

Shell surface coarsely reticulate and spinose; short, stubby spines arranged in haphazard way at about valve centre (Pl. 24, fig. 7) along ridges of reticulae. Towards valve margins, however, reticulation and spines become aligned in rows parallel to valve margins (Pl. 27, figs. 1, 6, 9). This linear arrangement often produces concentric appearance.

Internally, hinge hemimerodont; terminal elements only weakly dentate/loculate. No accommodation groove.

Inner margin and line of concrescence coincide to produce broad anterior duplicature. Anteriorly, 14 straight radial pore canals (text-fig. 42B). Selvage occupies peripheral position in both valves.

Muscle scars consist of slightly oblique row of 4 oval adductor scars with oval antero-dorsal frontal scar (text-fig. 42A).

Dimensions (in mm)

	Length	Height	Width
Holotype, Io.4669, right valve	0·52	0·32	
Paratype, Io.4670, left valve	0·54	0·33	
,, Io.4671, left valve	0·54	0·34	
,, Io.4672, carapace	0·45	0·31	0·20
,, Io.4673, left valve	0·52	0·29	
,, Io.4674, right valve	0·49	0·31	
,, Io.4645, left valve	0·54	0·34	
,, Io.4676, left valve	0·52	0·34	
,, Io.4677, left valve	0·51	0·32	
,, Io.4678, left valve	0·44	0·28	

Remarks. H. imitata, named after its external similarity to *Echinocythereis* Puri (1953), is a common species in the Santonian but does not appear to be particularly common in the Campanian as far as the samples examined are concerned.

Cythere harrisiana reticosa Jones and Hinde, described from the Gingin Chalk by Chapman (1917), is not synonymous with the species described by Jones and Hinde (1890) from the Gault of England. It is, however, conspecific with *Hystrico-cythere imitata,* Chapman's figured specimen (CPC.7136) being a juvenile carapace.

Although there is as yet insufficient material to do more than suggest that the Campanian specimens are more quadrate in outline than the Santonian forms, the specimen illustrated (Pl. 27, fig. 9) would certainly support this point of view.

INCERTAE SEDIS

Genus A sp.

Plate 9, figs. 8–13

Remarks. 3 specimens of this small new genus have been found in the Toolonga Calcilutite, Sample 4 (Santonian). In the absence of additional material, needed to provide information concerning the muscle scars and radial pore canals, assignment to a known family is not possible.

Externally the species represented here is uniquely ornamented and the presence of sexual dimorphism is indicated by the existence of elongate males (Pl. 9, figs. 8, 13) and more quadrate females (Pl. 9, figs. 9–12). Internally the hinge appears to be amphidont, but the state of preservation is not good.

The male dimorph bears some resemblance ornamentally to *Cythereis quadri-latera* of Chapman (1917), but differences in the carapace outline suggest that the 2 are not even congeneric.

Dimensions (in mm)

	Length	Height	Width
Io.4590, female carapace	0·37	0·20	0·19
Io.4591, female carapace	0·37	0·21	0·18
Io.4592, male left valve	0·40	0·21	

Genus B sp.

Plate 20, fig. 11; Plate 21, figs. 1–4; Plate 22, figs. 7, 8

Remarks. This genus probably belongs to the Cytherideidae, but the absence of muscle scar detail makes the assignment uncertain.

10 specimens of this rather weak shelled species have been found, of which 4 have been obtained from the Korojon Calcarenite, Sample 1 (Campanian), and 6 from the Toolonga Calcilutite, Sample 4 (Santonian).

Internally the hinge appears to be lophodont and 6 long, straight radial pore canals have been seen to cross the broad anterior duplicature.

Dimensions (in mm)

	Length	*Height*
Io.4621, right valve	0·51	0·25
Io.4623, right valve	0·56	0·27
Io.4624, left valve	0·49	0·26
Io.4625, left valve	0·49	0·26
Io.4626, right valve	0·48	0·24

SUMMARY

The Upper Cretaceous sediments of Coniacian, Santonian, and Campanian age obtained from cores from the Carnarvon Basin, Western Australia, have yielded a rich ostracod fauna of which 11 genera and 54 species are identified as being new. It has been possible, however, to name only 9 new genera and 32 new species, principally owing to a lack of material in some of the forms.

Much work has been done on the Cretaceous ostracods of Europe, North America, South America, Asia, and Africa whilst Australasia has been largely neglected. As a result, it has been possible to recognize 8 new genera in Western Australia which are not represented in the Northern Hemisphere. A ninth, *Oculocytheropteron*, may be shown eventually to be a southern form, but at this moment the genus includes one northern species (p. 49). 3 genera, *Rostrocytheridea* Dingle (1969a), *Majungaella* Grekoff (1963), and *Cytheralison* Hornibrook (1952), have been recorded only from the southern hemisphere and so far only from the circum-Indian Ocean region. Of these, *Cytheralison* is perhaps more restricted in that it has not been recorded from outside the Australasian area, either fossil or as a living genus. *Rostrocytheridea* is present in the Neocomian of South Africa, whilst *Majungaella* is known to have a range of Callovian to Campanian. Indeed, this genus occupies, in the Southern Hemisphere, the stratigraphical position of 2 Northern Hemisphere genera, namely, *Progonocythere* Sylvester-Bradley (1948) in the Jurassic and *Neocythere* Mertens (1956) in the Cretaceous.

Some of the new genera are considered to be ancestral to more cosmopolitan Tertiary and Recent forms; thus, *Eorotundracythere* may be an early relative of the living Australasian genus *Rotundracythere* Mandelstam (1958), and *Hermanites sagitta* appears to be a close ancestor of *Hermanites* Puri (1955) from the Miocene of Florida. *Paramunseyella* and *Premunseyella* are both regarded as ancestral to the Tertiary to Recent *Munseyella* Bold (1957) and *Pectocythere* Hanai (1957). McKenzie's suggestion (1967, p. 229) that *Munseyella* arose in the Gulf Coast Region of North America must be reconsidered in the light of a possible eastward migration from Australasia of both *Munseyella* and *Pectocythere*. Indeed, the Australasian region may eventually be shown, as already suggested by Hornibrook

(1952) and McKenzie (1967), to have been the development area for a number of ostracod genera besides those just mentioned. For example, *Trachyleberis* Brady (1898), common in the Tertiary and now world-wide in distribution, is here recorded from Coniacian to Campanian sediments. *Oculocytheropteron*, with 1 possible living Northern Hemisphere species, *O. nodosum* (Brady 1868), is ancestral to at least 10 species currently living in the Southern Hemisphere, of which 8 live in Australasian seas at the present time. The remaining genera are cosmopolitan and possibly have their origins in the Northern Hemisphere. Of these, *Cytherella* Jones (1849) is unique in that 3 species have been shown to possess 2 internal, posterior brood pouches in the female dimorph. The normally cited number is 1 brood cavity for *Cytherella* and 2 for *Cytherelloidea* Alexander (1929), but as far as southern representatives are concerned, this simple distinction can no longer be made.

Homeomorphy has been observed in 3 Australian genera in external carapace shape and ornamentation, but the internal characters in all instances are clearly distinctive. The most striking example is of *Apateloschizocythere*, which externally is virtually indistinguishable from the homeomorphic genera *Schizocythere* Triebel (1950) and *Cnestocythere* Triebel (1950). All 3 genera are considered to be related and are placed in the family Schizocytheridae. Homeomorphy amongst unrelated ostracods is evidenced by the very close similarity of *Hystricocythere* to *Echinocythereis* Puri (1955) and of *Toolongella mimica* to *Mauritsina hieroglyphica* (Bosquet 1847), the type species of that genus (see Deroo 1962).

The Santonian interval can be clearly identified from the Campanian sediments above (Table 2), whilst the paucity of the Coniacian fauna makes the distinction of this horizon even more obvious. The boundary between the Santonian and the Campanian lies within the Toolonga Calcilutite. The Campanian Korojon Calcarenite is correlatable with the upper part of the Toolonga Calcilutite (see Table 1). 12 ostracod species range through the Santonian/Campanian interval, whilst 8 are limited to the Santonian and 28 restricted to the Campanian.

Transgression of the Australian continent during Upper Cretaceous times was never extensive and the sediments (calcarenites, calcilutites, shales, and siltstones) are essentially of shallow water deposition. Across the Indian Ocean, the beds of equivalent age in East Africa are deep water shales in which ostracods are rare and quite dissimilar (Bate 1969). In South Africa sandstones and sandy limestones appear and the environment of deposition is much closer to that of Western Australia.

In West Africa the Upper Cretaceous ostracod fauna (Reyment 1960) has *Buntonia*, *Protobuntonia*, *Brachycythere*, *Ovocytheridea*, and *Veenia* dominant. These genera are continued into North Africa (Bold 1964; Oertli 1966; Salahi 1966) and India (unpublished material). In South Africa the above-mentioned ostracods are absent, apart from *Brachycythere*, of which 3 species are recorded by Dingle (1969*b*). This situation is also true for East Africa where only 1 species of *Brachycythere* has been recorded (Bate 1969) and this from the Turonian. The Western Australian Upper Cretaceous does not contain any of these ostracods but has a link with East and South Africa by the presence of the 2 ostracod genera *Majungaella* and *Rostrocytheridea*. On this basis it is possible to identify a Tethyan faunal province which incorporates West and North Africa and India and a Southern Hemisphere province incorporating East and South Africa and Australia. The presence of *Brachycythere*

| CONIACIAN | | SANTONIAN | CAMPANIAN | | | STAGES |
UPPER GEARLE SILTSTONE SAMPLE No.6	UPPER GEARLE SILTSTONE SAMPLE No.5	TOOLONGA CALCILUTITE SAMPLE No.4	TOOLONGA CALCILUTITE SAMPLE No.3	KOROJON CALCARENITE SAMPLE No.2	KOROJON CALCARENITE SAMPLE No.1	CORE SAMPLES / OSTRACODS
●						Trachyleberis sp. Type A.
	●					Cytherella sp. Type C.
	●					Trachyleberidea sp.
	●					Trachyleberis sp. Type B.
		●				Cytherella sp. Type A.
		●				? Bythocypris sp. Type A.
		●				Paramunseyella austracretacea
		●				Premunseyella ornata
		●				Eorotundracythere compta
		●				Rostrocytheridea canaliculata
		●				Cythereis brevicosta
		●				Scepticocythereis ornata
		●				? Toolongella sp.
		●				Genus A sp.
		●			●	Cytherella sp. Type B.
		●	●	●		Cytherelloidea cobberi
		●	●	●	●	Cytherelloidea westaustraliensis
		●	●	●		Majungaella annula
		●	●	●	●	Apateloschizocythere geniculata
		●	●	●	●	Cytheropteron (Cytheropteron) carinoalatum
		●	●	●	●	Oculocytheropteron praenuntatum
		●	●	●	●	Hermanites sagitta
		●	●	●	●	Karsteneis (Karsteneis) aspericava
		●	●	●	●	Toolongella mimica
		●	●			Hystricocythere imitata
		●			●	Genus B sp.
			●			Cytherella atypica
			●			Platella sp.
			●			Paracypris sp.
			●			Pontocyprella sp.
			●			Krithe sp.
			●			Pedicythere sp.
				●		Cytherella alata
				●	●	Cytherelloidea carnarvonensis
			●	●	●	Bairdia austracretacea
			●	●		Pontocyprella dorsoconvexa
				●		Bythoceratina sp.
			●	●		Cytheralison contorta
				●		Monoceratina invenusta
				●	●	Premunseyella imperfecta
			●	●	●	Eorotundracythere levigata
				●		Paracytheridea sp.
			●	●		Anebocythereis amoena
				●	●	Curfsina levigata
			●	●	●	Limburgina formosa
			●	●	●	Oertliella exquisita
			●	●		Trachyleberis anteplana
					●	Cytherella sp. Type D.
					●	Cytherella sp. Type E.
					●	? Bythocypris sp. Type B.
					●	Cytherura sp.
					●	Cytheropteron (Infracytheropteron) anotum
					●	? Orthonotacythere sp.
					●	Costa elongata

TABLE 2. Range table of the Upper Cretaceous Ostracoda, Carnarvon Basin.

in East and South Africa has already been suggested (Bate 1969) as being indicative of a sea connection with West Africa and South America.

In terms of specific correlation with Upper Cretaceous sediments outside Western Australia, this has only been observed in unpublished material examined from Papua, where the ostracod *Cytherelloidea westaustraliensis* was identified.

Accepting that the continents have drifted apart since the Jurassic, the Indian Ocean would have been very much smaller in the Cretaceous than at the present time. Maps showing the position of Australia either close to East Africa or abutting against the east coast of India during the Cretaceous are unlikely to be correct, as evidenced by the ostracod faunas. Indeed, the only close correlation observed has been with Papua, suggesting that Australia was closely associated with this region and with the S.E. Asia block during the Cretaceous.

The uniqueness of the Australian fauna further suggests that Australia was separated from Africa by rather more than just a sea connection. It is considered, therefore, that the continental block of Antarctica was situated between Africa and Australia.

The considerable degree of exploration currently being initiated by oil companies in S.E. Asia should bring to light further information concerning the geographical position of Australia during the Cretaceous and add to our knowledge concerning the ostracod faunas of that time.

REFERENCES

ALEXANDER, C. I. 1927. The stratigraphic range of the Cretaceous ostracod *Bairdia subdeltoidea* and its allies. *J. Paleont.*, Sharon, **1**, 29–33, pl. 6.

—— 1929. The Ostracoda of the Cretaceous of North Texas. *Bull. Univ. Texas*, **2907**, 1–137, pls. 1–10.

BATE, R. H. 1963. Middle Jurassic Ostracoda from North Lincolnshire. *Bull. Br. Mus. nat. Hist. (Geol.)*, London, **8**, 173–219, pls. 1–15.

—— 1967. *In* HARLAND, W. B. *et al.* (eds.) *The Fossil Record*, London (Geological Society), xii+828 pp.

—— 1969. *In* BATE, R. H. and BAYLISS, D. D. An outline account of the Cretaceous and Tertiary Foraminifera and of the Cretaceous ostracods of Tanzania. *3rd African Micropalaeont. Coll.*, Cairo, 113–164, 8 pls.

BELFORD, D. J. 1959. Stratigraphy and micropalaeontology of the Upper Cretaceous of Western Australia. *Geol. Rdsh.*, Stuttgart, **47**, 629–647.

—— 1960. Upper Cretaceous Foraminifera from the Toolonga Calcilutite and Gingin Chalk, Western Australia. *Bull. Bureau Min. Res.*, Canberra, **57**, 1–198, 35 pls.

BOLD, W. A. VAN DEN. 1946. *Contribution to the Study of Ostracoda.* 167 pp. 18 pls. Amsterdam.

—— 1957. Ostracoda from the Paleocene of Trinidad, *Micropaleont.*, New York, **3**, 1–18, pls. 1–4.

—— 1964. Ostracoden aus der Oberkreide von Abu Rawash, Ägypten. *Palaeontographica*, Stuttgart, **123**, 111–136, pls. 13–15.

BONNEMA, J. H. 1941. Ostracoden aus der Kreide des Untergrundes der nordöstlichen Niederlande. *Natuurh. Maandbl.*, Maastricht, **29** and **30**, 27 pp.

BRADY, G. S. 1868. A monograph of the Recent British Ostracoda. *Trans. Linn. Soc.*, London, **26**, 353–495, pls. 23–41.

—— 1880. Report on the Ostracoda dredged by H.M.S. Challenger during the years 1873–1876. *Challenger Exped.*, 1873–1876, *Rept., Zool.*, London, **1**, 1–179, pls. 1–44.

BROWN, D. A. *et al.* 1968. *The Geological Evolution of Australia and New Zealand*, x+409 pp. Oxford.

BUTLER, E. A. and JONES, D. E. 1957. Cretaceous Ostracoda of Prothro and Rayburns Salt Domes, Bienville Parish, Louisiana. *Bull. geol. Surv. La.*, Baton Rouge, **32**, 1–65, 6 pls.

CHAPMAN, F. 1917. Monograph of the Foraminifera and Ostracoda of the Gingin Chalk. *Bull. geol. Surv. West. Aust.*, Perth, **72**, 1-87, pls. 1-14.

—— and SHERBORN, C. D. 1893. On the Ostracoda of the Gault at Folkestone. *Geol. Mag.*, London, **10**, 345-349, pl. 14.

CONDON, M. A. 1968. The Geology of the Carnarvon Basin, Western Australia. Pt. **3**: Post Permian Stratigraphy; Structure; Economic Geology. *Bull. Bureau Min. Res.*, Canberra, **77**, 1-68.

—— *et al.* 1956. The Giralia and Marrilla Anticlines North West Division, Western Australia. *Bull. Bureau Min. Res.*, Melbourne, **25**, 1-86, pls. 1-5.

CORYELL, H. N. and FIELDS, S. 1937. A gatun ostracode fauna from Cativa, Panama. *Am. Mus. Novit.*, New York, **956**, 1-18.

——, SAMPLE, C. H., and JENNINGS, P. H. 1935. *Bairdoppilata*, a new genus of Ostracoda with two new species. *Am. Mus. Novit.*, New York, **777**, 1-5.

CRANE, M. 1965. Upper Cretaceous ostracodes of the Gulf Coast area. *Micropaleont.*, New York, **11**, 191-254, pls. 1-9.

DEROO, G. 1962. Mauritsininae, nouvelle sous-famille de Cytheridae (Ostracodes) dans le Crétacé Supérieur de la région de Maastricht, Pays-Bas. *Rev. Micropaléont.*, Paris, **4**, 203-210, pls. 1, 2.

—— 1966. Cytheracea (Ostracodes) du Maastrichtien de Maastricht (Pays-Bas) et des régions voisines; résultats stratigraphiques et paléontologiques de leur étude. *Meded. geol. Sticht.*, Maastricht, **2**, 1-197, 27 pls.

DINGLE, R. V. 1969*a*. Marine Neocomian Ostracoda from South Africa. *Trans. R. Soc. S. Afr.*, Cape Town, **38**, 139-163, pl. 9.

—— 1969*b*. Upper Senonian ostracods from the coast of Pondoland, South Africa. Ibid. 347-385.

DONZE, P. 1962. Contribution à l'étude paléontologique de l'Oxfordien Supérieur de Trept (Isère); 111-Ostracodes. *Trav. Lab. Géol. Fac. Sc. Lyon*, **8**, 125-142, pls. 9-11.

—— 1964. Ostracodes Berriasiens des Massifs Subalpins Septentrionaux (Bauges et Chartreuse). *Trav. Lab. Géol. Fac. Sc. Lyon*, **11**, 103-160, pls. 1-9.

EAGAR, S. 1965. Ostracoda of the London Clay (Ypresian) in the London Basin: **1**, Reading District. *Rev. Micropaléont.*, Paris, **8**, 15-32, pls. 1-2.

EDGELL, H. S. 1954. The stratigraphical value of *Bolivinoides* in the Upper Cretaceous of north-west Australia. *Contr. Cushman Fdn.*, Sharon, **5**, 68-76.

—— 1957. The genus *Globotruncana* in Northwest Australia. *Micropaleont.*, New York, **3**, 101-126, pls. 1-4.

ELOFSON, O. 1941. Zur Kenntnis der marinen Ostracoden Schwedens, mit besonderer Berücksichtigung des Skagerraks. *Uppsala Univ. Zool. Bidr.* **19**, 215-534.

GREKOFF, N. 1963. Contribution à l'étude des Ostracodes du Mésozoïque Moyen (Bathonien-Valanginien) du Bassin de Majunga, Madagascar. *Rev. Inst. franç. Pétrole*, Paris, **18**, 1709-1762, pls. 1-10.

GRÜNDEL, J. 1964. Neue Ostracoden aus der deutschen Unterkreide **II**. *Mber. dt. Akad. Wiss. Berl.* **6**, 849-858, pls. 1, 2.

HANAI, T. 1957. Studies on the Ostracoda from Japan, **II**. Subfamily Pectocytherinae n. subfam. *J. Fac. Sci. Univ. Tokyo*, **10**, 469-482, pl. 11.

—— 1970. Studies on the ostracod subfamily Schizocytherinae Mandelstam. *J. Paleont.*, Tulsa, **44**, 693-729, pls. 107, 108.

HAZEL, J. E. 1965. Notes on the ostracode genus *Trachyleberidea* Bowen. Ibid. **39**, 501-503.

—— 1967*a*. *Repandocosta*, a new Cretaceous and Tertiary ostracode genus. Ibid. **41**, 103-110, pls. 15, 16.

—— 1967*b*. Classification and distribution of the Recent Hemicytheridae and Trachyleberididae (Ostracoda) off north-eastern North America. *Prof. Pap. U.S. geol. Surv.*, Washington, **564**, 1-49, pls. 1-11.

HERRIG, E. 1966. Ostracoden aus der Weissen Schreibkreide (Unter-Maastricht) der Insel Rügen. *Paläont. Abh. A. Paläozool.*, Berlin, **11**, 693-1024, pls. 1-45.

HIRSCHMANN, N. 1916. Ostracoda of the Baltic Sea collected by N. M. Knipovitch and S. A. Pavlovitch in the summer of 1908. *Zool. Mus. Akad. Nauk*, Leningrad, **20**, 569-597 [in Russian].

HOLDEN, J. C. 1964. Upper Cretaceous ostracods from California. *Palaeontology*, **7**, 393-429.

HORNIBROOK, N. DE B. 1952. Tertiary and Recent marine Ostracoda of New Zealand. *Bull. N. Z. geol. Surv.*, Wellington, **18**, 1-82, pls. 1-18.

HOWE, H. V. 1934. The ostracode genus *Cytherelloidea* in the Gulf Coast Tertiary. *J. Paleont.*, Menasha, **8**, 29-34, pl. 5.

HOWE, R. C. 1963. Type saline Bayou Ostracoda of Louisiana. *Bull. geol. Surv. La.*, Baton Rouge, **40**, 1–62, pls. 1–4.

JONES, T. R. 1849. A monograph of the Entomostraca of the Cretaceous Formation of England. *Palaeontogr. Soc. (Monogr.)*, London, 40 pp., pls. 1–7.

—— and HINDE, G. J. 1890. A supplementary monograph of the Cretaceous Ostracoda of England and Ireland. *Palaeontogr. Soc. (Monogr.)*, London, i–viii+70 pp., pls. 1–4.

KAYE, P. 1963a. The ostracod genus *Neocythere* in the Speeton Clay. *Paleontology*, **6**, 274–281, pl. 41.

—— 1963b. Ostracoda of the subfamilies of Protocytherinae and Trachyleberidinae from the British Lower Cretaceous. *Paläont. Z.*, Stuttgart, **37**, 225–238.

—— 1964. Ostracoda of the genera *Eucytherura* and *Cytheropteron* from the Speeton Clay. *Geol. Mag.*, London, **101**, 97–107, pls. 4, 5.

—— and BARKER, D. 1965. Ostracoda from the Sutterby Marl (U. Aptian) of South Lincolnshire. *Palaeontology*, **8**, 375–390, pls. 48–50.

KEIJ, A. A. 1957. Eocene and Oligocene Ostracods of Belgium. *Mem. Inst. r. Sci. nat. Belg.*, Bruxelles, **136**, 1–210, 23 pls.

KINGMA, J. T. 1948. *Contributions to the knowledge of the Young-Caenozoic Ostracoda from the Malayan region.* 106 pp., 11 pls. Utrecht.

KRÖMMELBEIN, K. 1967. Ostracoden aus der marinen 'Küsten-Kreide' Brasiliens. 2, *Sergipella transatlantica* n.g., n.sp., und *Aracajuia benderi* n.g., n.sp., aus dem Ober-Aptium/Albium. *Senck. leth.*, Frankfurt a.M., **48**, 525–533, pl. 1.

LOETTERLE, G. J. 1937. The micropaleontology of the Niobrara Formation in Kansas, Nebraska, and South Dakota. *Neb. geol. Surv. (Bull.)*, Lincoln, **12**, 1–73, pls. 1–11.

MADDOCKS, R. F. 1969. Revision of Recent Bairdiidae (Ostracoda). *Bull. U.S. natn. Mus.*, Washington, **295**, 1–126.

MANDELSTAM, M. I., in ABUSHIK, A. F. *et al.* 1958. New genera and species of Ostracods. Microfauna of the U.S.S.R. *Trud. vses. neft. nauch.-issled. geol. Inst. (VNIGRI)*, Leningrad, **9**, 232–287 [in Russian].

—— and SHNEIDER, G. F. The fossil Ostracoda of the USSR. Family Cyprididae. Ibid. **203**, 1–242, 42 pls.

MCKENZIE, K. G. 1967. The distribution of Caenozoic marine Ostracoda from the Gulf of Mexico to Australia. *Publs. Syst. Ass.*, London, **7**, 219–238.

MCWHAE, J. R. H. *et al.* 1958. The stratigraphy of Western Australia. *J. geol. Soc. Aust.*, Adelaide, **4**, 1–161.

MERTENS, E. 1956. Zur grenzziehung Alb/Cenoman in Nordwesdeutschland mit Hilfe von Ostracoden. *Geol. Jb.*, Hannover, **72**, 173–230, pls. 8–14.

MOORE, R. C. 1961. (Ed.) *Treatise on Invertebrate Paleontology*, Pt. Q. Arthropoda, **3**, xxiii+442 pp., 334 figs. Univ. Kansas Press.

MOOS, B. 1966. Die Ostracoden-Fauna des Unteroligozäns von Bünde (Bl. Herford-West, 3817) und einige verwandte Arten aus verschiedenen Tertiärstufen (Ostr., Crust.). *Geol. Jb.*, Hannover, **84**, 281–298, 2 pls.

MÜLLER, G. W. 1908. Die Ostracoden der Deutschen Südpolar-Expedition 1901–1903. *Dt. Sudpol. Exped.*, Berlin, **10**, 51–81, pls. 4–19.

NEALE, J. W. 1960. Marine Lower Cretaceous Ostracoda from Yorkshire, England. *Micropaleont.*, New York, **6**, 203–224, pls. 1–4.

—— 1962. Ostracoda from the type Speeton Clay (Lower Cretaceous) of Yorkshire. Ibid. **8**, 425–484, pls. 1–13.

—— 1967. An ostracod fauna from Halley Bay, Coats Land, British Antarctic Territory. *Br. Antarct. Surv. Sci. Rept.*, London, **58**, 1–50, pls. 1–4.

OERTLI, H. J. 1957. Ostracodes du Jurassique Supérieur du Bassin de Paris (Sondage Vernon 1). *Rev. Inst. franç. Pétrole*, Paris, **12**, 647–695, pls. 1–7.

—— 1958. Les ostracodes de l'Aptien-Albien D'Apt. *Rev. Inst. franç. Pétrole*, Paris, **13**, 1499–1537, pls. 1–9.

—— 1959. *Euryitycythere* und *Parexophthalmocythere*, zwei neue Ostrakoden-Gattungen aus der Unterkreide Westeuropas. *Paläont. Z.*, Stuttgart, **33**, 241–246, pl. 32.

—— 1966. Étude des ostracodes du Crétacé Supérieur du Bassin Côtier de Tarfaya. *Notes et Mém. Serv. Géol.*, Pau, **175**, 267–278, pls. 1, 2.

POKORNY, V. 1963. *Karsteneis* gen.n. (Ostracoda, Crustacea) from the Upper Cretaceous of Bohemia. *Cas. Miner. Geol.*, Praha, **8**, 39–44.

POKORNY, V. 1964. *Oertliella* and *Spinicythereis*, new ostracod genera from the Upper Cretaceous. *Vest. ústred. Ust. geol.*, Praha, **39**, 283–284, 1 pl.

PURI, H. S. 1953. Contribution to the study of the Miocene of the Florida Panhandle; Pt. **III**—Ostracoda. *Geol. Bull. Fla.*, Tallahassee, **36**, 218–345, pls. 1–17.

—— 1955. *Hermanites*, new name for *Hermania* Puri 1954. *J. Paleont.*, Tulsa, **29**, 558.

—— 1960. Recent Ostracoda from the west coast of Florida. *Trans. Gulf Coast Assoc. geol. Soc.*, Jackson, **10**, 107–149, pls. 1–6.

RAMSAY, W. V. 1968. A new morphological aspect of the ostracode genus *Cytherelloidea* Alexander. *Micropaleont*, New York, **14**, 348–356.

REYMENT, R. A. 1960. Studies on Nigerian Upper Cretaceous and Lower Tertiary Ostracoda. I, Senonian and Maestrichtian Ostracoda. *Acta Univ. Stockholm*, **7**, 1–238, pls. 1–23.

RUGGIERI, G. 1962. Gli ostracodi marini del Tortoniano (Miocene medio superiore) di Enna, nella Sicilia centrale. *Palaeontogr. ital.*, Pisa, **56** (N.S. **26**), 1–68+I–VI, pls. 1–7.

SALAHI, D. 1966. Ostracodes du Crétacé supérieur et du Tertiaire en provenance d'un sondage de la région de Zelten (libye). *Rev. Inst. franç. Pétrole*, Paris, **21**, 3–42, pls. 1–5.

SARS, G. O. 1866. Oversigt af Norges marine ostracoder. *Vidensk.-Selsk. i Christiania, Forh.*, Oslo, **7**, 1–130.

SOHN, I. G. 1960. Paleozoic species of *Bairdia* and related genera. *Prof. Pap. U.S. geol. Surv.*, Washington, **330A**, 1–105, pls. 1–6.

SYLVESTER-BRADLEY, P. C. 1948. Bathonian ostracods from the Boueti Bed of Langton Herring, Dorset. *Geol. Mag.* London, **85**, 185–204, pls. 12–15.

—— 1950. The shell of the ostracod genus *Bairdia*. *Ann. Mag. nat. Hist.*, London, **3**, 751–756.

TRIEBEL, E. 1950. Homöomorphe Ostracoden-Gattungen. *Senckenbergiana*, Frankfurt a.M., **31**, 313–330, pls. 1–4.

—— 1957. Neue Ostracoden aus dem Pleistozän von Kalifornien. *Senck. leth.*, Frankfurt a.M., **38**, 291–309.

—— 1958. Zwei neue Ostracoden-Gattungen aus dem Lutet des Pariser Beckens. Ibid. **39**, 105–117.

VAN MORKHOVEN, F. P. C. M. 1963. *Post Paleozoic Ostracoda*, **2**, 478 pp. Amsterdam.

VAN VEEN, J. E. 1934. Die Cypridae und Bairdiidae der Maastrichter Tuffkreide und des Kunrader Korallenkalkes von Süd-Limburg. *Natuurh. Maandbl.*, Maastricht, **23**, 88–132, pls. 1–8.

—— 1936. Die Cytheridae der Maastrichter Tuffkreide und des Kunrader Korallenkalkes von Süd-Limburg. *Natuurh. Maandbl.*, Maastricht, **25**, 131–170, 9 pls.

WAGNER, C. W. 1957. Sur les Ostracodes du Quaternaire Récent des Pays-Bas et leur utilisation dans l'étude géologique des dépôts holocènes. *Diss. Univ. Paris*, 259 pp., 50 pls.

R. H. BATE
Department of Palaeontology
British Museum (Natural History)
Cromwell Road
London, S.W.7

Revised typescript received 26 May 1971

PLATES

EXPLANATION OF PLATE 1

Figs. 1–6. *Cytherelloidea westaustraliensis* sp. nov. Campanian. p. 10

1, Female right valve, paratype Io.4404, ×80.
2, Female right valve, holotype Io.4397, ×79.
3, Female left valve, paratype Io.4403, ×75.
4, Male right valve, paratype Io.4398, ×75.
5, Internal view of male right valve to show muscle scars and peripheral contact groove; paratype Io.4401, × 77.
6, Juvenile left valve; this specimen was not given a metallic coating prior to investigation in the scanning electron microscope; the photograph was taken at a reduced voltage of 3·5 kv; paratype Io.4436, ×100.

Figs. 7, 8. *Cytherelloidea cobberi* sp. nov. Campanian. p. 9

7, Left valve, paratype Io.4387, ×77.
8, Right valve, holotype Io.4386, ×75.

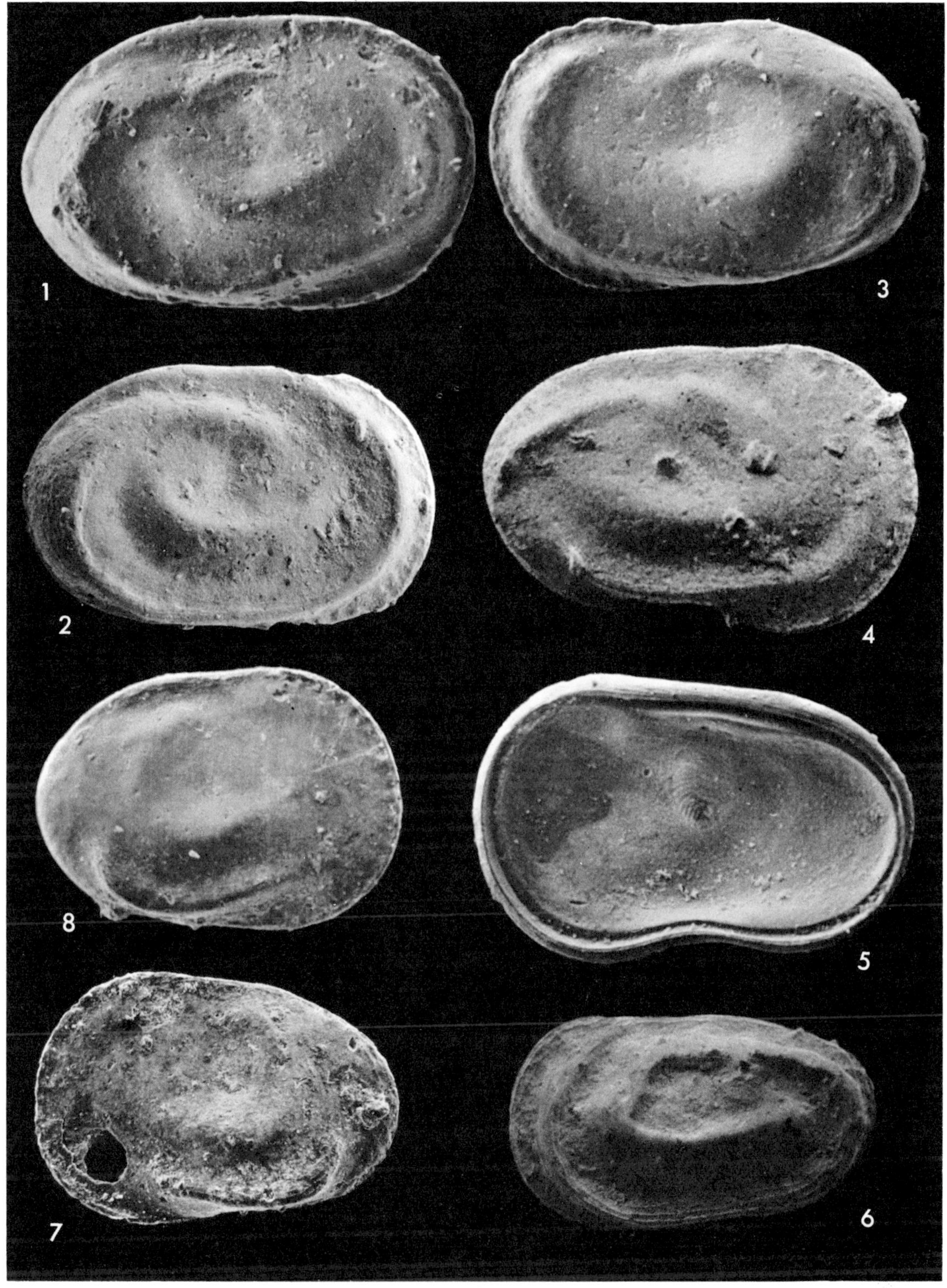

BATE, Upper Cretaceous Ostracoda, Carnarvon Basin

EXPLANATION OF PLATE 2

Figs. 1, 2. *Platella* sp. Campanian. p. 14

1, Left valve, Io.4419, $\times$140.
2, Left valve, Io.4418, $\times$120.

Fig. 3. *Cytherelloidea westaustraliensis* sp. nov. Campanian. p. 10

Dorsal view of female carapace, paratype Io.4402, $\times$75.

Figs. 4–9. *Cytherelloidea carnarvonensis* sp. nov. Campanian. p. 13

4, Dorsal view, female carapace; paratype Io.4414, $\times$100.
5, Male right valve, paratype Io.4412, $\times$100.
6, Left side, male carapace; holotype Io.4410, $\times$100.
7, Female left valve, paratype Io.4413, $\times$100.
8, Right side, female carapace; paratype Io.4411, $\times$100.
9, Internal view, female right valve; paratype Io.4415, $\times$100.

BATE, Upper Cretaceous Ostracoda, Carnarvon Basin

BATE, Upper Cretaceous Ostracoda, Carnarvon Basin

EXPLANATION OF PLATE 4

Figs. 1, 2, 5, 8, 11, 12. *Bairdia austracretacea* sp. nov. Campanian. p. 16

1, Rosette of 8 muscle scars with 2 small antero-ventral scars situated beneath; right valve, paratype Io.4423, × 500.

2, Internal view to show position of muscle scars and details of hinge and duplicature; right valve, paratype Io.4423, × 85.

5, Left valve, paratype Io.4421, × 50.

8, Left valve, paratype Io.4424, × 50.

11, Left valve, paratype Io.4428, × 50.

12, Right valve, paratype Io.4425, × 50.

Figs. 3, 4, 9. *Pedicythere* sp. Campanian. p. 52

Right, left, and dorsal views, complete carapace, Io.4497, × 100.

Fig. 6. *Pontocyprella* sp. Campanian. p. 21

Internal view, left valve, Io.4437, × 50.

Figs. 7, 10. *Pontocyprella dorsoconvexa* sp. nov. Campanian. p. 19

7, Left valve, paratype Io.4439, × 50.

10, Right valve, holotype Io.4440, × 50.

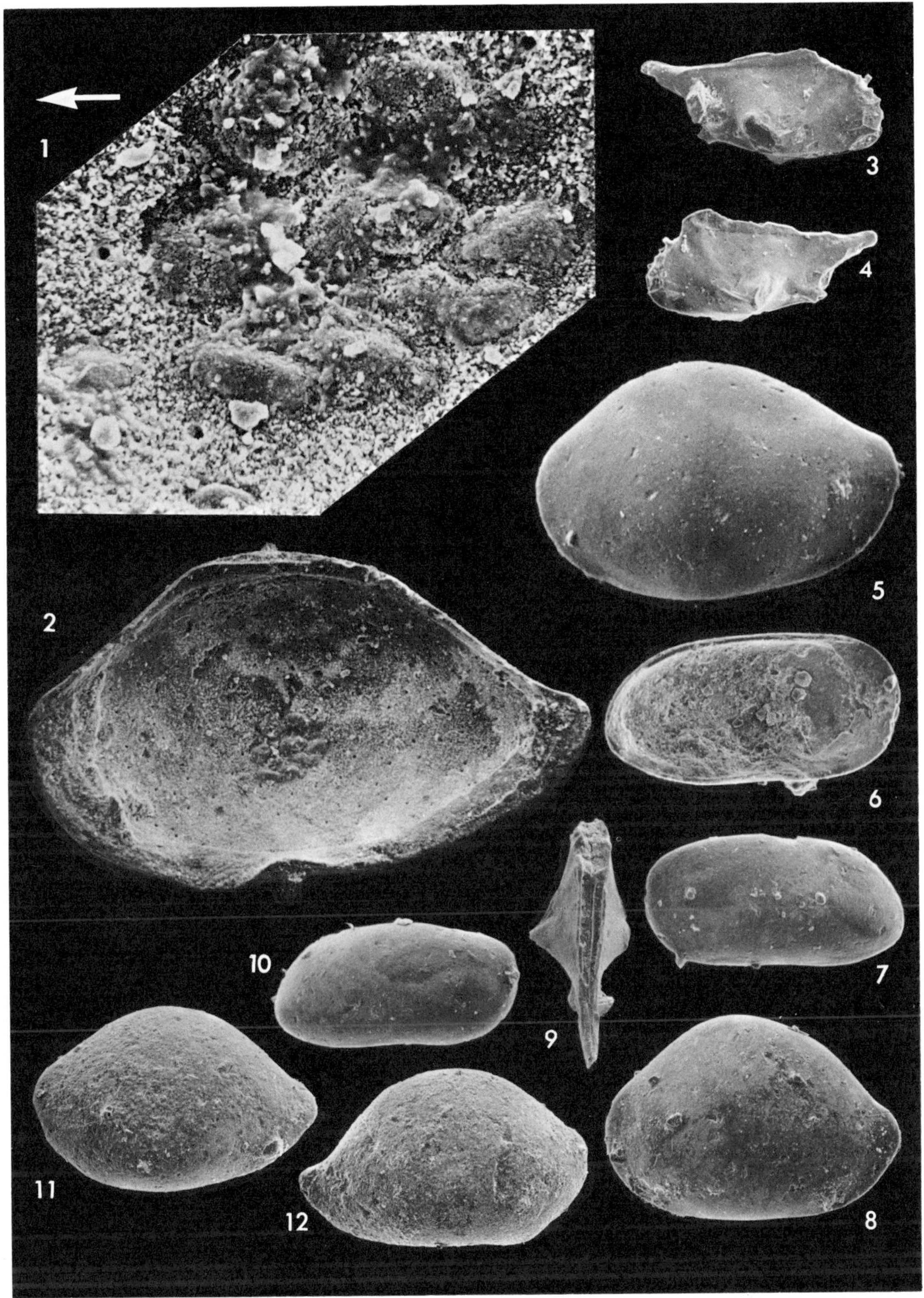

BATE, Upper Cretaceous Ostracoda, Carnarvon Basin

EXPLANATION OF PLATE 5

BATE, Upper Cretaceous Ostracoda, Carnarvon Basin

EXPLANATION OF PLATE 6

Fig. 1. ?*Orthonotacythere* sp. Campanian. p. 52

Right valve, Io.4496, × 100.

Figs. 2–7. *Majungaella annula* sp. nov. Santonian, Campanian. p. 26

2, Left valve, paratype Io.4455, Santonian, × 100.

3, Right side of complete carapace, holotype Io.4450, Santonian, × 100.

4, Dorsal view, paratype Io.4453, Santonian, × 100.

5, Muscle scars, showing four adductor scars with large, oval, antero-dorsal frontal scar; a small scar is situated on the anterior side of the frontal scar; paratype Io.4459, Campanian, × 700.

6, Hinge, left valve, paratype Io.4459, Campanian, × 145.

7, Hinge, right valve, paratype Io.4463, Campanian, × 145.

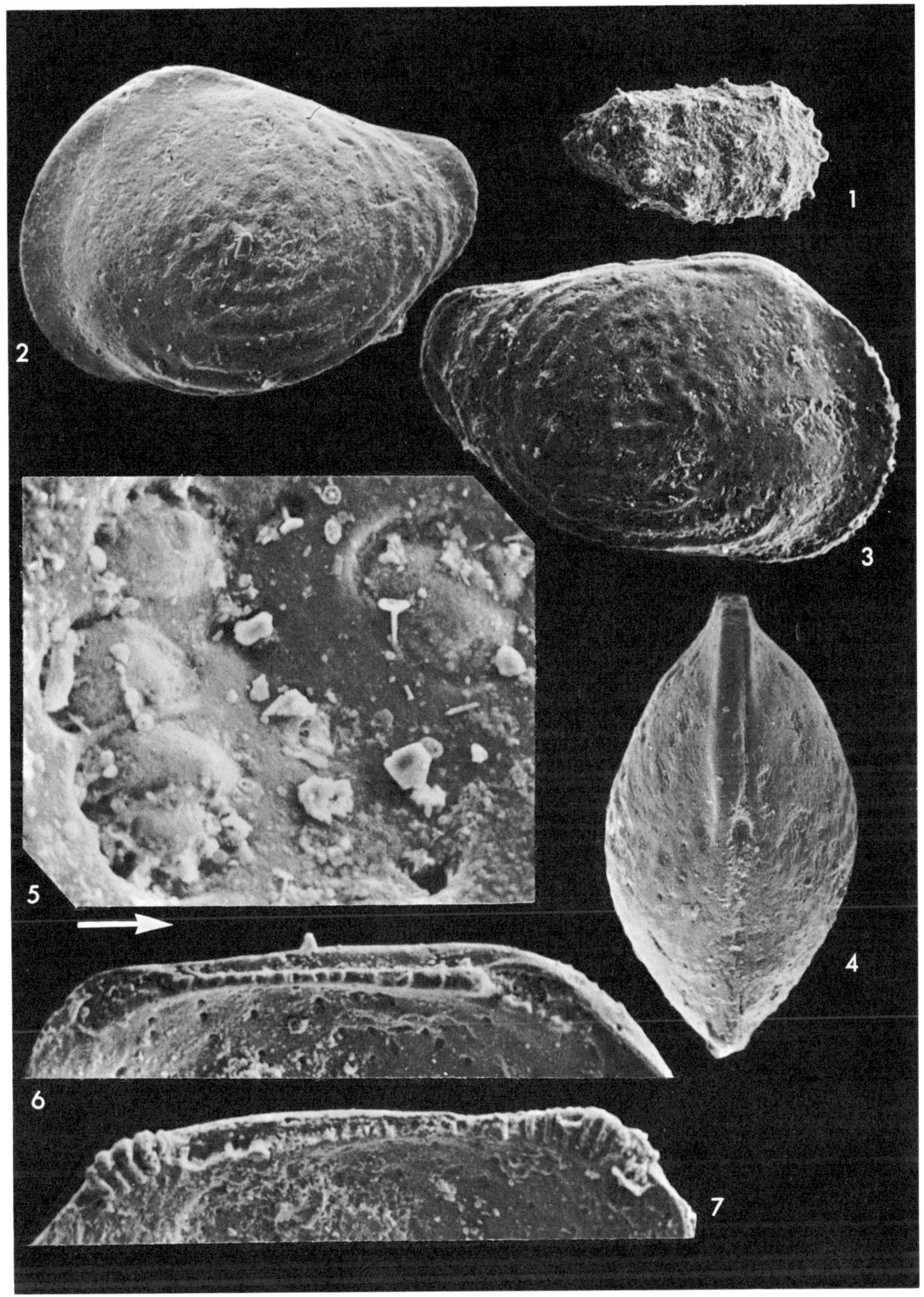

BATE, Upper Cretaceous Ostracoda, Carnarvon Basin

EXPLANATION OF PLATE 7

Figs. 1–4. *Majungaella annula* sp. nov. Campanian. p. 26

1, Left valve, paratype Io.4461, ×95.
2, Right valve, paratype Io.4462, ×100.
3, Internal view, left valve, paratype Io.4459, ×100.
4, Internal view, right valve, paratype Io.4460, ×100.

Figs. 5–8. *Apateloschizocythere geniculata* gen. et sp. nov. Campanian. p. 30

5, Right valve, paratype Io.4471, ×100.
6, Left valve, paratype Io.4467, ×100.
7, Right valve, paratype Io.4466, ×100.
8, Internal view, left valve, paratype Io.4478, ×100.

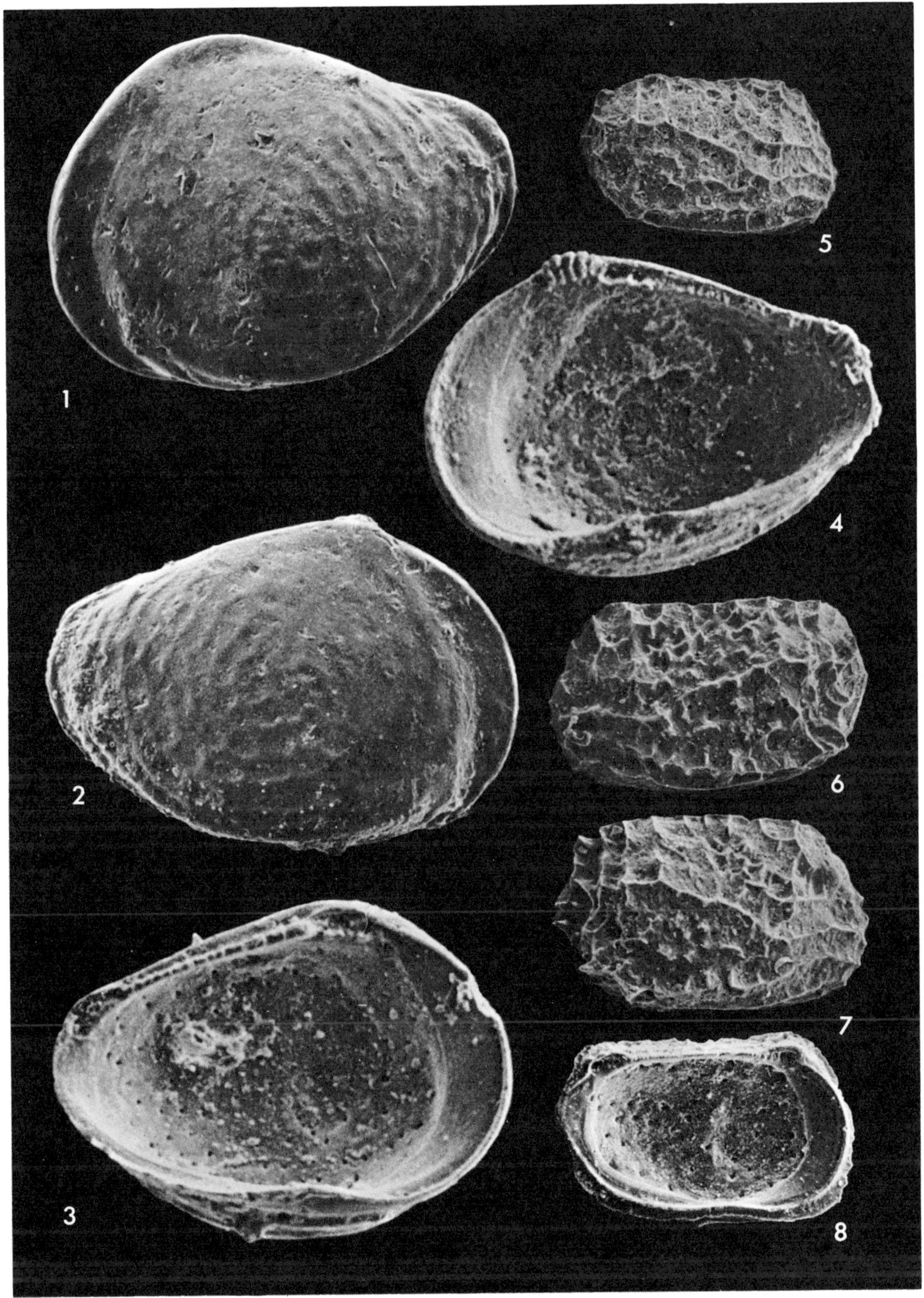

BATE, Upper Cretaceous Ostracoda, Carnarvon Basin

EXPLANATION OF PLATE 8

Figs. 1–10. *Apateloschizocythere geniculata* gen. et sp. nov. Campanian. p. 30

1, Dorsal view, carapace, paratype Io.4473, × 100.

2, 3, Right valve hinge to show anterior and posterior terminal teeth and loculate median groove; paratype Io.4470, × 600.

4, Ventral view, carapace, paratype Io.4477, × 100.

5, Left valve, paratype Io.4472, × 100.

6, Left valve, holotype Io.4465, × 100.

7, Left valve, paratype Io.4467, × 100.

8, 9, Left valve hinge to show posterior and anterior terminal loculate sockets and part of the dentate median bar; paratype Io.4478, × 600.

10, Internal view, right valve, paratype Io.4470, × 100.

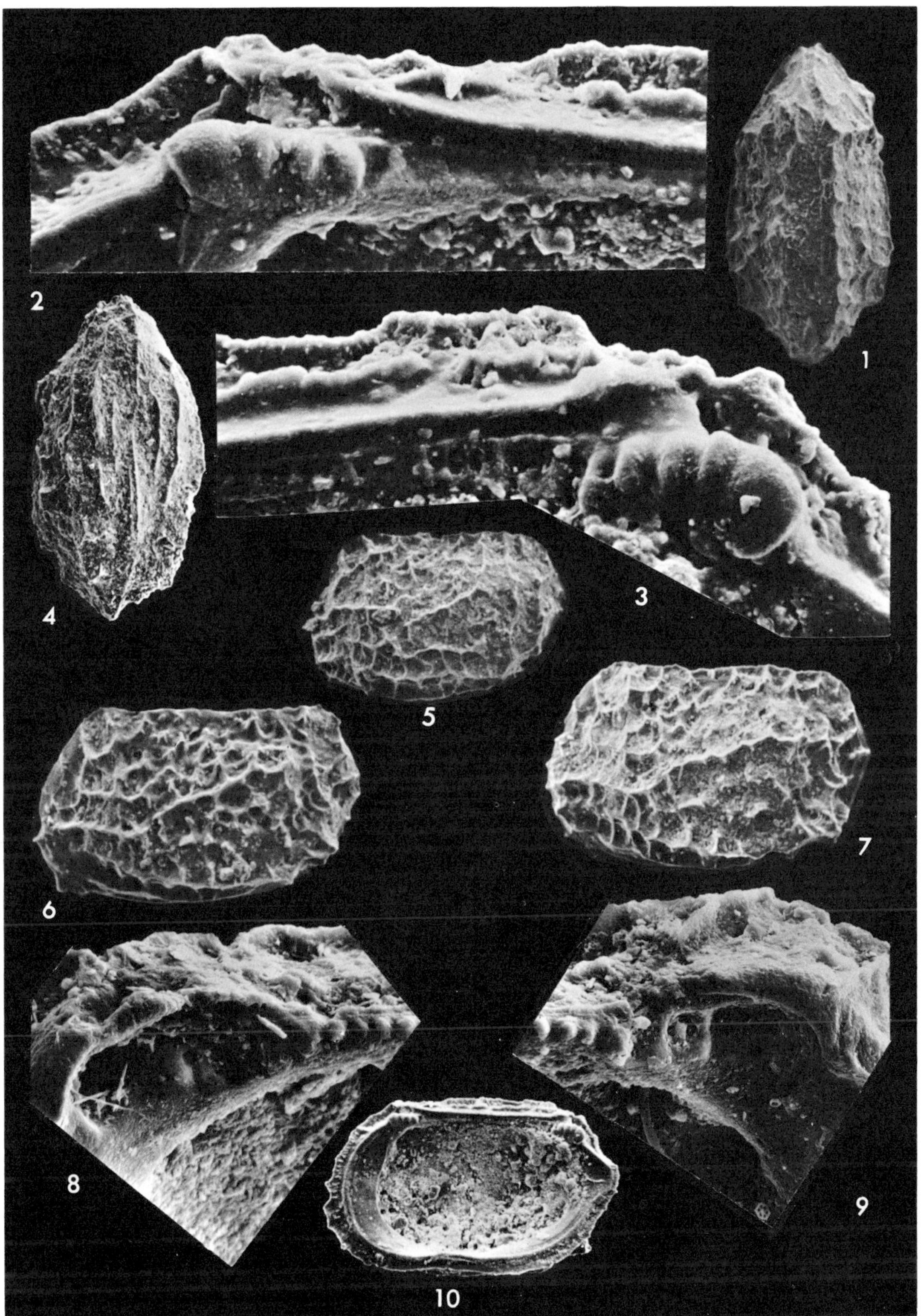

BATE, Upper Cretaceous Ostracoda, Carnarvon Basin

EXPLANATION OF PLATE 9

Figs. 1–4. *Karsteneis* (*Karsteneis*) *aspericava* sp. nov. Santonian, Campanian.   p. 61
   1, Ventral view, female carapace, paratype Io.4493, Santonian, ×95.
   2, Dorsal view, female carapace, holotype Io.4481, Campanian, × 85.
   3, Female right valve, paratype Io.4482, Campanian, × 85.
   4, Female left valve, paratype Io.4483, Campanian, ×85.

Figs. 5–7. *Premunseyella imperfecta* gen. et sp. nov. Campanian.   p. 36
   5, Right side of male carapace, holotype Io.4514, × 100.
   6, Female left valve, paratype Io.4516, × 100.
   7, Dorsal view of female carapace, paratype Io.4515, × 100.

Figs. 8–13. Genus A sp. Santonian.   p. 78
   8, 13, External and internal views, male left valve, Io.4592, × 100.
   9, 11, Left and dorsal views, female carapace, Io.4591, × 100.
   10, 12, Ventral and right views, female carapace, Io.4590, × 100.

BATE, Upper Cretaceous Ostracoda, Carnarvon Basin

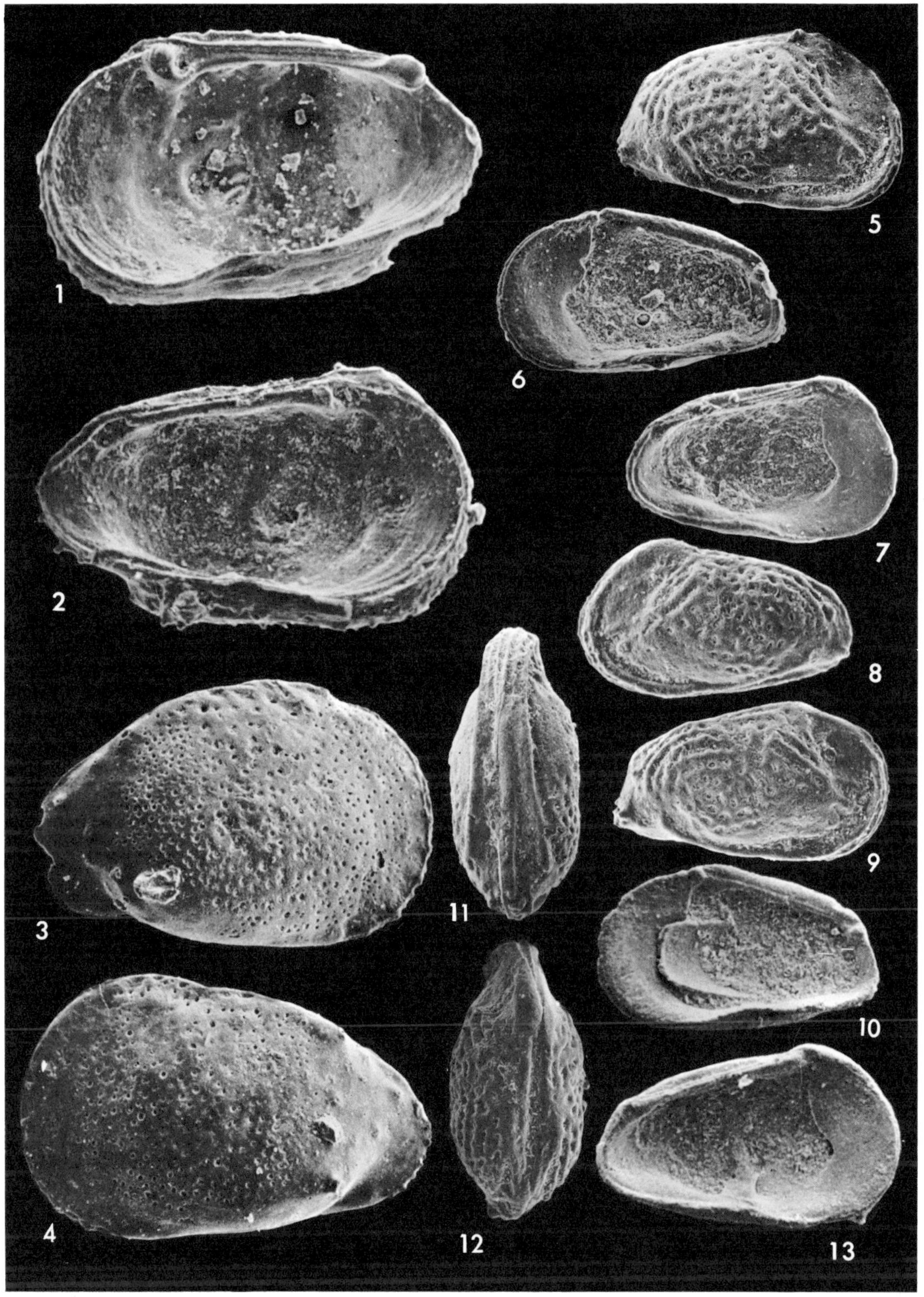

BATE, Upper Cretaceous Ostracoda, Carnarvon Basin

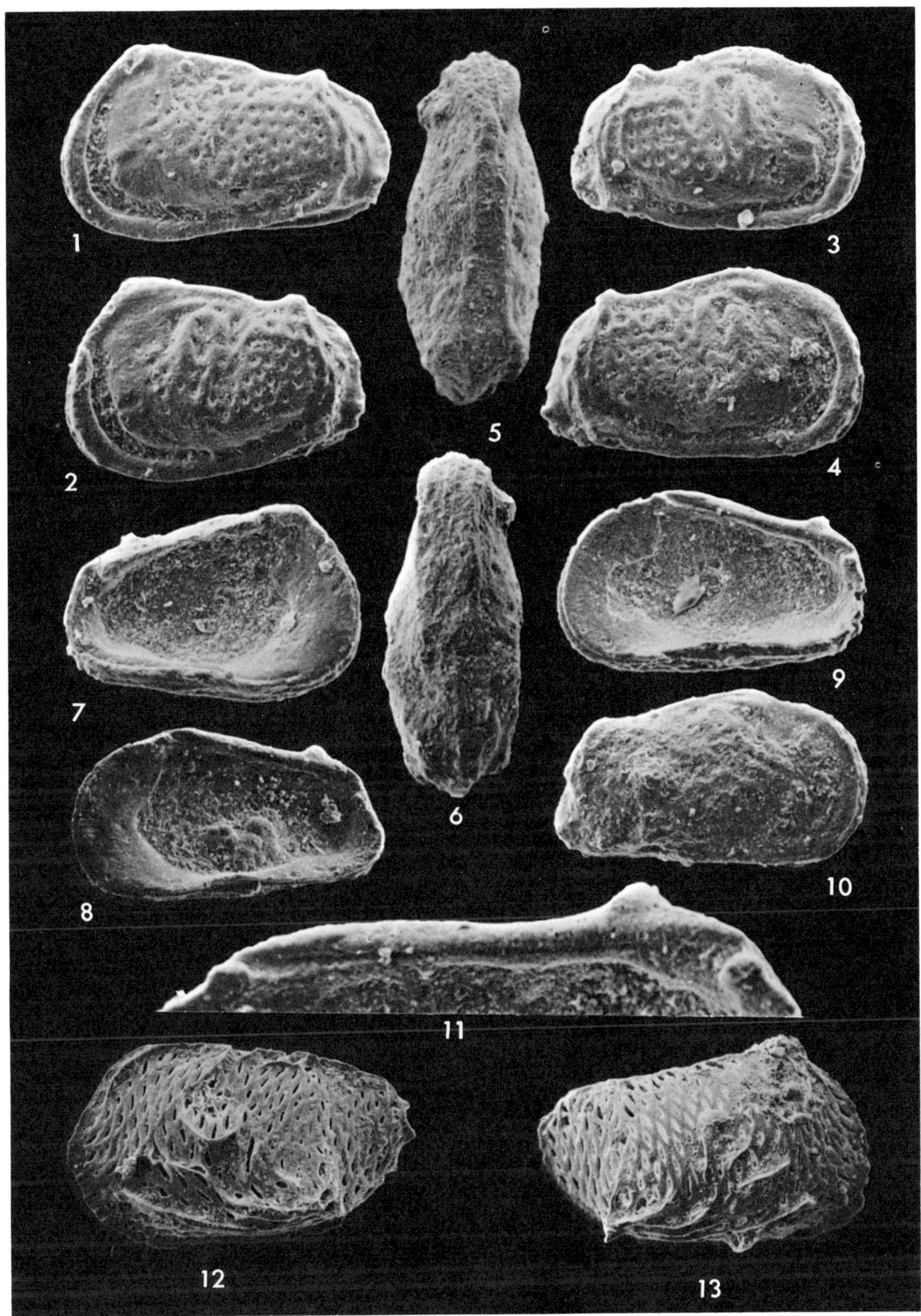

BATE, Upper Cretaceous Ostracoda, Carnarvon Basin

EXPLANATION OF PLATE 12

Figs. 1, 2, 4, 8. *Cytheralison contorta* sp. nov. Campanian.　　　　　　　p. 22
  1, Right valve, holotype Io.4521, × 100.
  2, Left valve, paratype Io.4522, × 100.
  4, Internal view, right valve, holotype Io.4521, × 100.
  8, Internal view, left valve, paratype Io.4526, × 100.

Figs. 3, 5, 6, 7. *Eorotundracythere compta* gen. et sp. nov. Santonian.　　　p. 42
  3, Female right valve, paratype Io.4551, × 100.
  5, Female left valve, holotype Io.4548, × 100.
  6, Internal view, female left valve, paratype Io.4552, × 100.
  7, Internal view, female right valve, paratype Io.4550, × 100.

Fig. 9. *Rostrocytheridea canaliculata* sp. nov. Santonian.　　　　　　　　p. 38
  Left valve, holotype Io.4532, × 100.

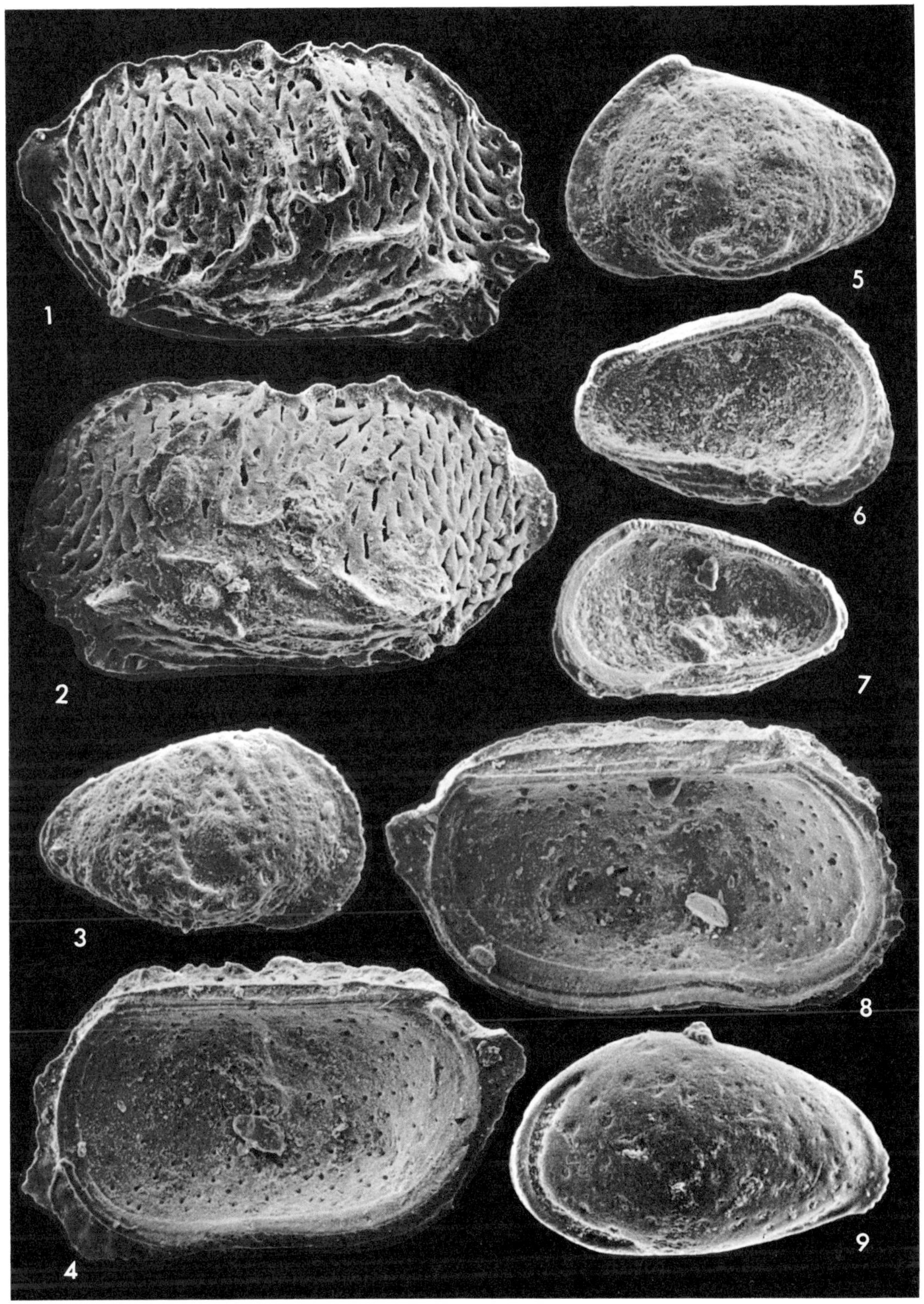

BATE, Upper Cretaceous Ostracoda, Carnarvon Basin

EXPLANATION OF PLATE 13

Figs. 1, 2, 5. *Anebocythereis amoena* gen. et sp. nov. Campanian. p. 53

1, Right valve, paratype Io.4559, × 100.
2, Internal view, right valve, paratype Io.4558, × 100.
5, Juvenile left valve, specimen lost, × 100.

Fig. 3. *Cytheralison contorta* sp. nov. Campanian. p. 22

Left valve, paratype Io.4523, × 100.

Figs. 4, 6, 7, 8. *Rostrocytheridea canaliculata* sp. nov. Santonian. p. 38

4, Internal view, left valve, paratype Io.4535, × 100.
6, Right valve, paratype Io.4533, × 100.
7, Internal view, right valve, paratype Io.4536, × 100.
8, Enlarged view of right valve hinge, paratype Io.4536, × 240.

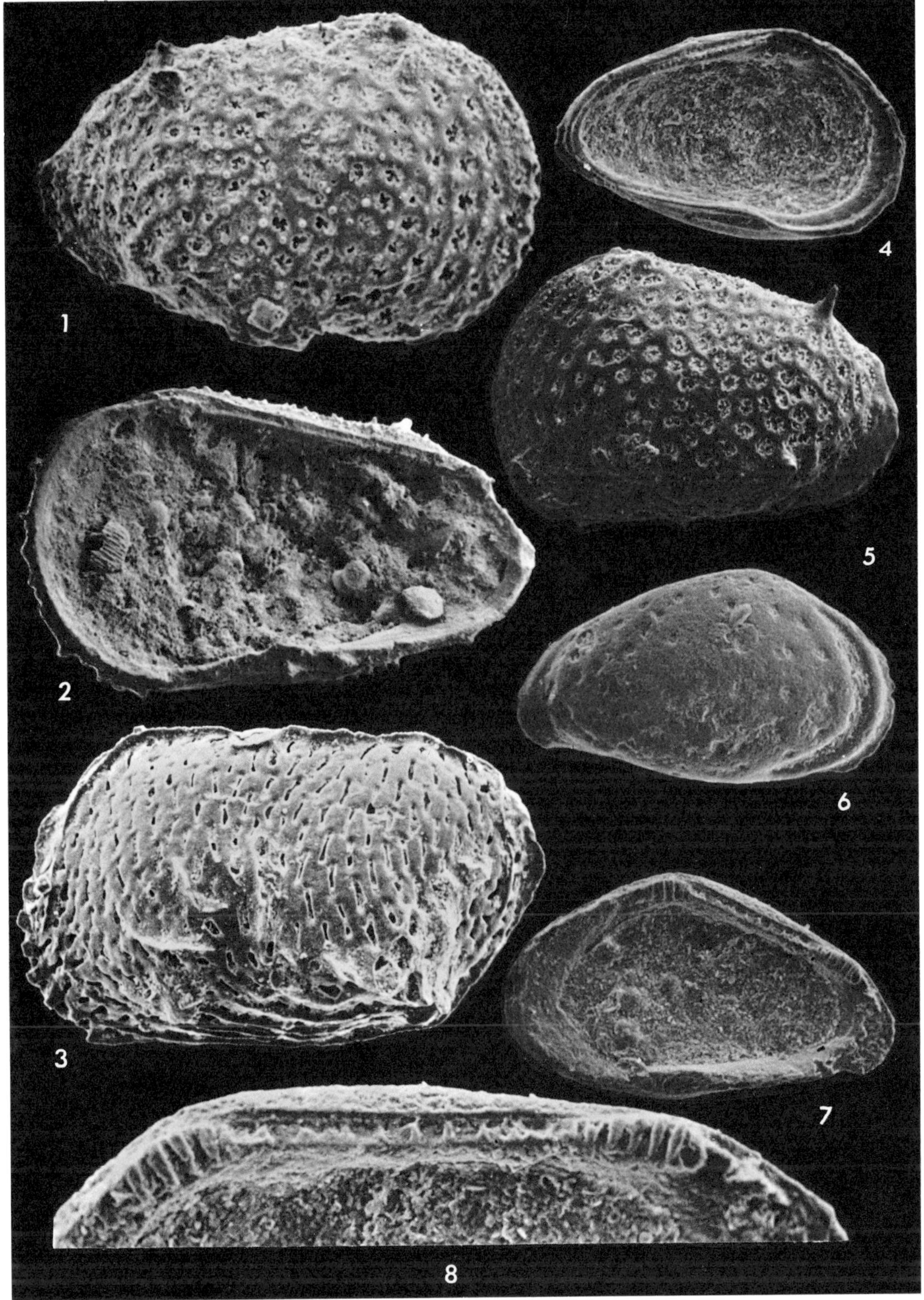

BATE, Upper Cretaceous Ostracoda, Carnarvon Basin

EXPLANATION OF PLATE 14

Figs. 1–3. *Anebocythereis amoena* gen. et sp. nov. Campanian. p. 53

1, Left valve, paratype Io.4557, × 100.
2, Left valve, holotype Io.4556, × 100.
3, Internal view, left valve, paratype Io.4561, × 100.

Figs. 4–8. *Eorotundracythere levigata* gen. et sp. nov. Campanian. p. 41

4, 8, Right and dorsal views, complete carapace, paratype Io.4540, × 100.
5, 7, Left and ventral views, complete carapace, holotype Io.4539, × 100.
6, Internal view, right valve, paratype Io.4541, × 100.

BATE, Upper Cretaceous Ostracoda, Carnarvon Basin

Figs. 1–3, 5, 6. *Anebocythereis amoena* gen. et sp. nov. Campanian.   p. 53

1, Central muscle scars: 4 oval adductor scars (add.), of which the lower 2 are in contact with each other whilst the upper 2 are separate. Antero-dorsal frontal scar(f.) and antero-ventral mandibular scar(m.); paratype Io.4560, ×620.

2, 3, Anterior (showing weak loculation) and posterior hinge sockets, left valve; paratype Io.4560, ×650.

5, Enlarged portion of ornament to show spine base and part of reticulation with small spines growing into pits. Spine base with apparent terminal pore, probably produced when spine tip is broken off; paratype Io.4559, ×1,000.

6, Enlarged portion of ornament to show more regular-shaped pits with triangular shaped, ingrowing spines; holotype Io.4556, × 540.

Fig. 4. *Karsteneis (Karsteneis) aspericava* sp. nov. Campanian.   p. 61

Normal pore canal aperture showing sieve plate; paratype Io.4482, ×2,000.

Fig. 7. *Apateloschizocythere geniculata* gen. et sp. nov. Campanian.   p. 30

Normal pore canal aperture showing sieve plate. Hp.-hair pore; Sp.-small pore; paratype Io.4679, ×2,000.

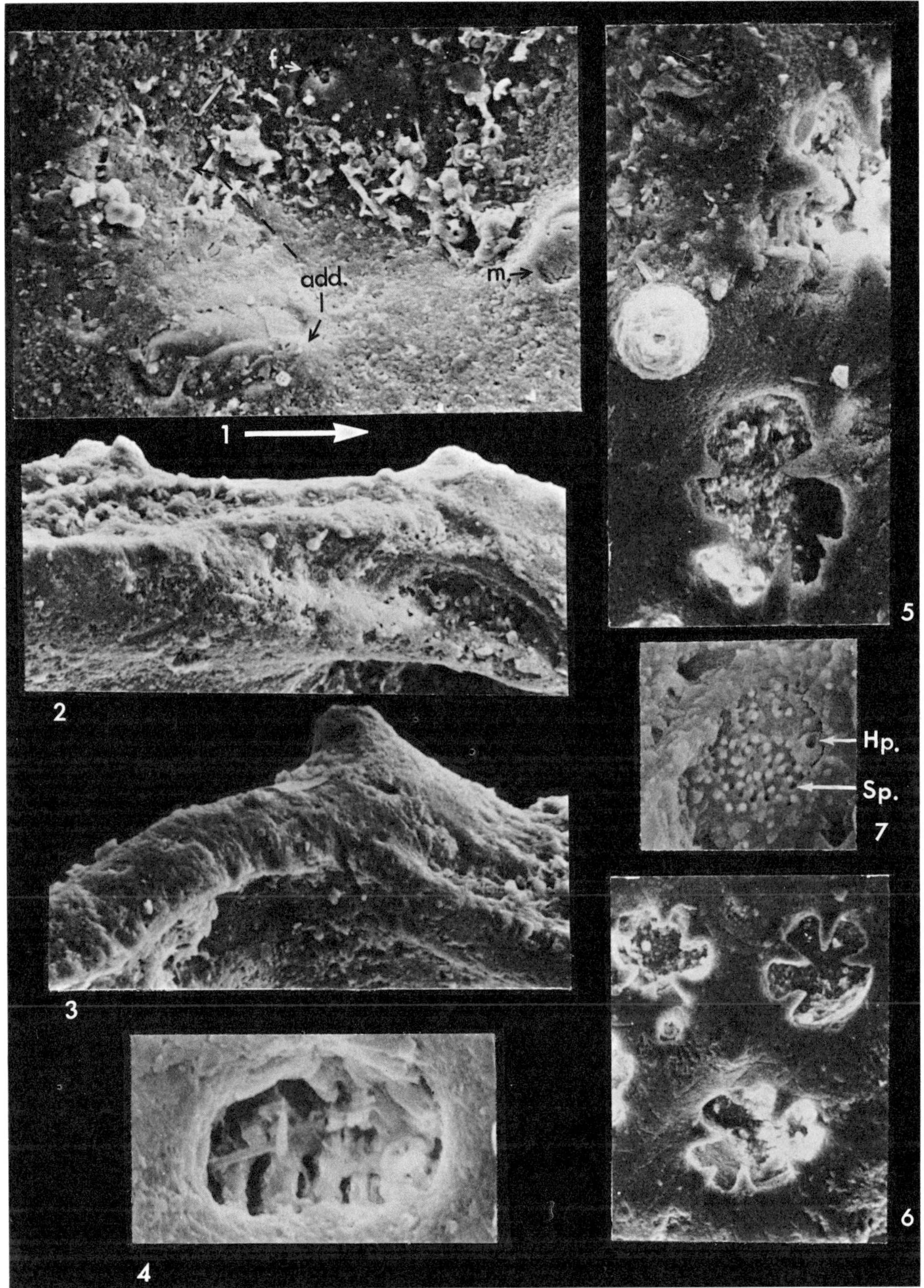

BATE, Upper Cretaceous Ostracoda, Carnarvon Basin

EXPLANATION OF PLATE 16

Figs. 1–12. *Oculocytheropteron praenuntatum* gen. et sp. nov. Campanian, Santonian. p. 50

1, Internal view, right valve, paratype Io.4586, Campanian, × 100.

2, 10, External and internal views, left valve; note duplicature, eye 'lens', muscle scars, and curved hinge in fig. 10; holotype Io.4581, Campanian, × 100.

3, Right valve, paratype Io.4585, Santonian, × 100.

4, Ventral view, complete carapace, paratype Io.4583, Campanian, × 100.

5, Dorsal view, left valve, paratype Io.4582, Campanian, × 100.

6, Enlarged portion of ornament in antero-dorsal region; note small punctae at anterior margin; holotype Io.4581, Campanian, × 550.

7, 8, Terminal loculate sockets and terminal dentition of median hinge bar, left valve; note ocular pit with eye 'lens' below anterior socket; holotype Io.4581, Campanian, × 1,200.

9, 11, 12, Internal view, right valve to show hinge with anterior and posterior elements enlarged (figs. 11, 12); paratype Io.4584, Santonian, × 100, × 1,280 and × 1,280 respectively.

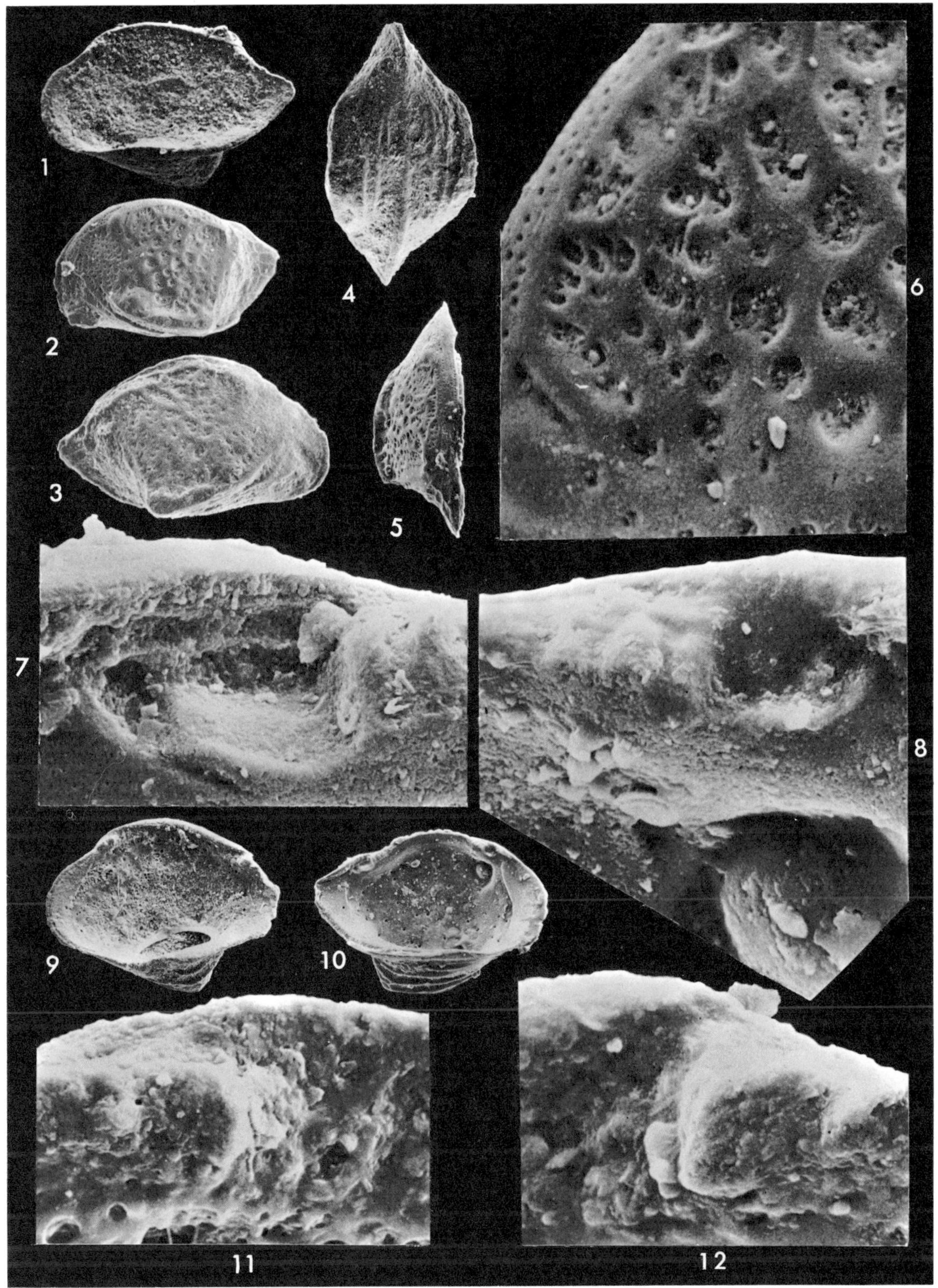

BATE, Upper Cretaceous Ostracoda, Carnarvon Basin

EXPLANATION OF PLATE 17

Figs. 1–9. *Cytheropteron* (*Infracytheropteron*) *anotum* sp. nov. Campanian. p. 48

1, 6, Right side and ventral view, complete carapace, holotype Io.4574, × 100.
2, Left valve, paratype Io.4577, × 100.
3, Right valve, paratype Io.4579, × 100.
4, Left valve, slightly more elongate in outline than fig. 2 and could represent a male dimorph; paratype Io.4576, × 100.
5, Dorsal view, complete carapace, paratype Io.4578, × 100.
7, Posterior hinge socket, left valve, paratype Io.4575, × 600.
8, 9, Internal view, to show terminal swellings of median hinge bar, and external view of left valve; paratype Io.4575, × 100.

Figs. 10, 11. *Cytheropteron* (*Cytheropteron*) *carinoalatum* sp. nov. Campanian. p. 45

10, Internal view, left valve, paratype Io.4567, × 100.
11, Left valve, holotype Io.4565, × 100.

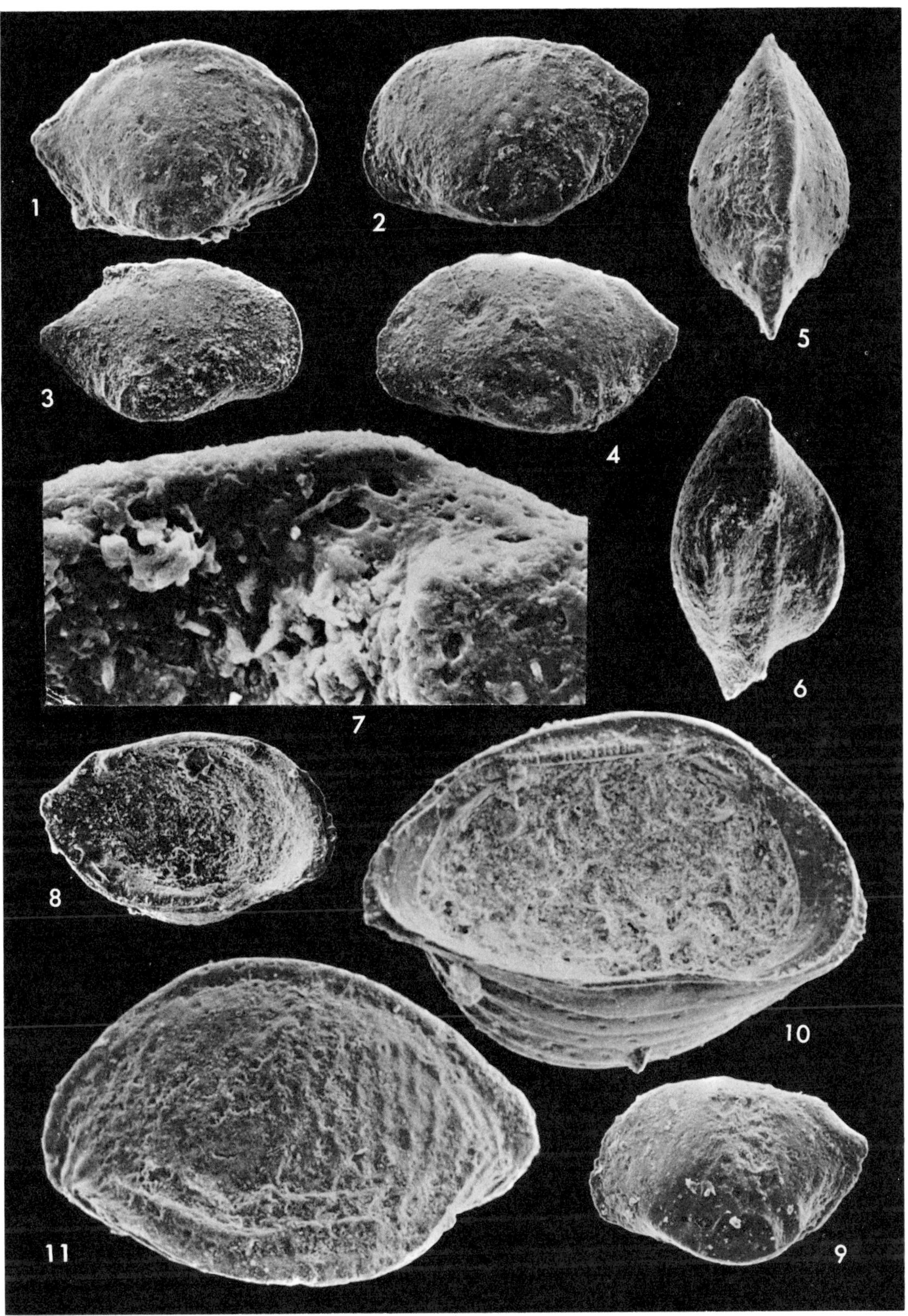

BATE, Upper Cretaceous Ostracoda, Carnarvon Basin

BATE, Upper Cretaceous Ostracoda, Carnarvon Basin

EXPLANATION OF PLATE 19

Fig. 1. *Cytheropteron* (*Cytheropteron*) *carinoalatum* sp. nov. Campanian. p. 45

Right valve hinge, paratype Io.4569, × 240.

Figs. 2, 3. *Oculocytheropteron praenuntatum* gen. et sp. nov. Campanian. p. 50

2, Enlarged median section of left valve hinge; paratype Io.4582, × 550.

3, Muscle scars (2 upper adductor scars and oral frontal scar), holotype Io.4581, ×245.

Fig. 4. *Cytheropteron* (*Infracytheropteron*) *anotum* sp. nov. Campanian. p. 48

Left valve hinge, paratype Io.4577, × 200.

Figs. 5, 8–10. *Toolongella mimica* gen. et sp. nov. Campanian, Santonian. p. 70

5, Left valve hinge, paratype Io.4598, Campanian, × 220.

8, Left valve, nodose variety, paratype Io.4601, Santonian, × 100.

9, Internal view, left valve, paratype Io.4599, Campanian, × 100.

10, Right valve, paratype Io.4597, Campanian, × 100.

Figs. 6, 7, 11. ?*Toolongella* sp. Santonian. p. 72

6, 7, Dorsal and ventral views, complete carapace, Io.4602, ×100.

11, Left valve, Io.4603, ×100.

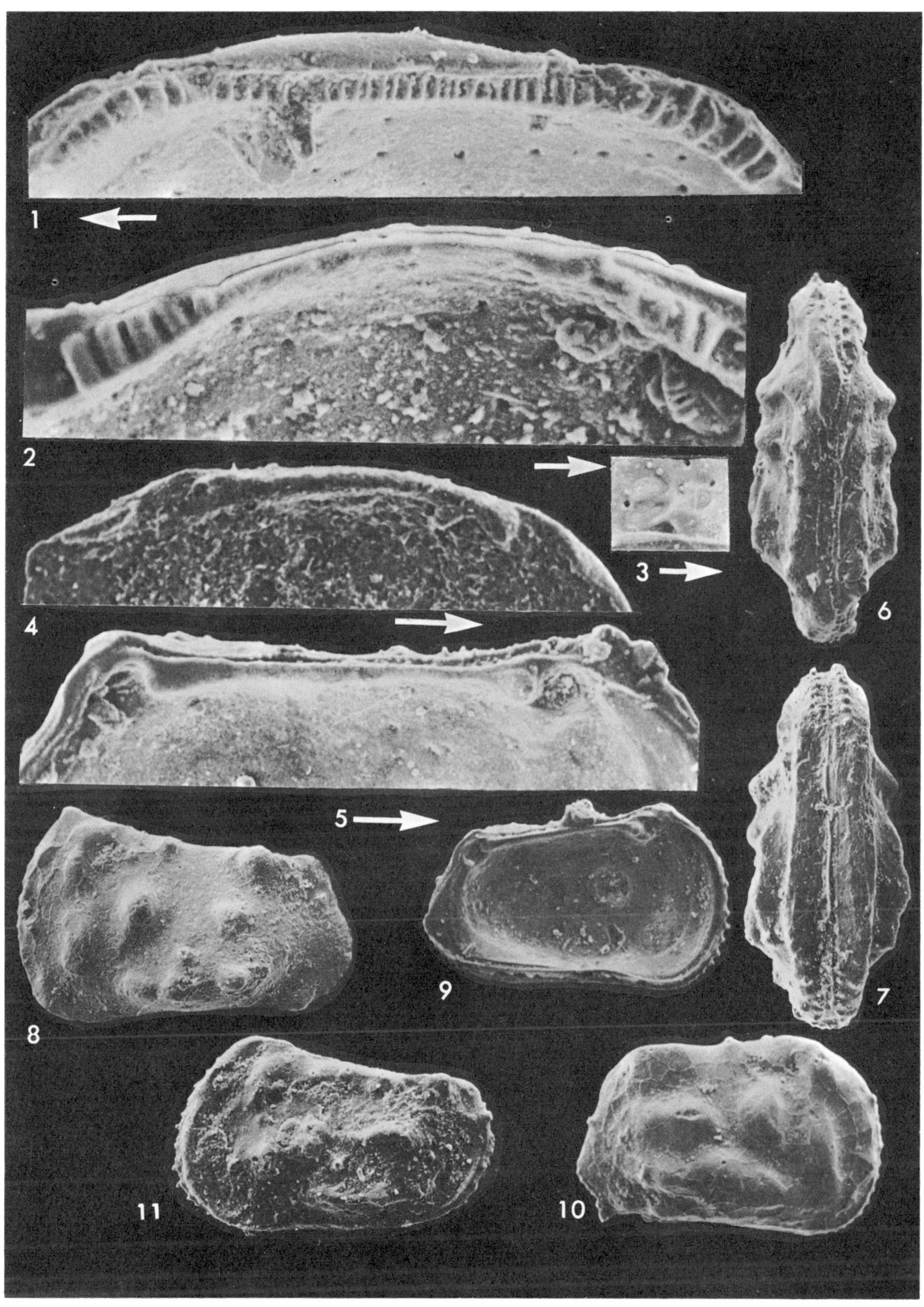

BATE, Upper Cretaceous Ostracoda, Carnarvon Basin

EXPLANATION OF PLATE 20

Figs. 1–4. *Toolongella mimica* gen. et sp. nov. Campanian. p. 70

1, Left valve, paratype Io.4595, ×100.
2, Internal view, left valve, paratype Io.4598, ×100.
3, Right valve, holotype Io.4593, ×100.
4, Internal view, right valve, paratype Io.4600, ×100.

Figs. 5–7. *Trachyleberis anteplana* sp. nov. Campanian. p. 74

5, Male left valve, paratype Io.4620, ×67.
6, Female right valve, holotype Io.4614, ×67.
7, Internal view, female right valve, paratype Io.4617, ×67.

Fig. 8. *Trachyleberis* sp. Type A. Coniacian. p. 74

Fragment of anterior margin, left valve, Io.4610, × 67.

Figs. 9, 10. *Cythereis brevicosta* sp. nov. Santonian. p. 57

9, Right valve, holotype Io.4605, ×62.
10, Left valve, paratype Io.4607, ×62.

Fig. 11. Genus B sp. Santonian. p. 78

Left valve hinge, Io.4624, × 240.

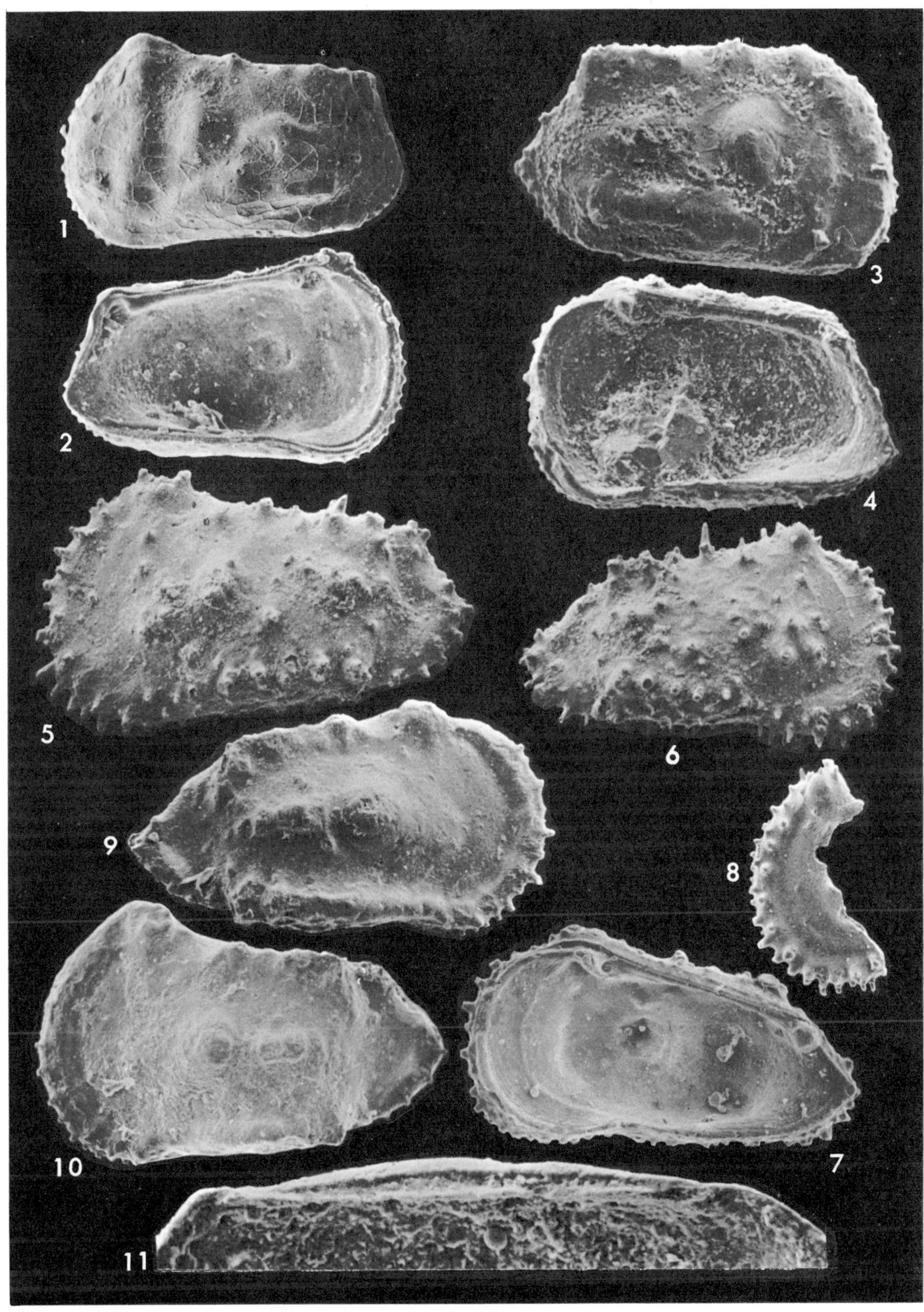

BATE, Upper Cretaceous Ostracoda, Carnarvon Basin

Figs. 1–4. Genus B sp. Campanian, Santonian.   p. 78

    1, Right valve, Io.4621, Campanian, ×100.
    2, Internal view, right valve, Io.4626, Santonian, ×100.
    3, Internal view, left valve, Io.4624, Santonian, ×100.
    4, Left valve, Io.4625, Santonian, ×100.

Figs. 5–7, 10–12. *Curfsina levigata* sp. nov. Campanian.   p. 56

    5, 10, Right side, female carapace ×100 and surface punctae ×1,200; paratype Io.4632.
    6, 11, Left side, female carapace ×100 and surface erosional pits ×600; holotype Io.4631.
    7, Male right valve, paratype Io.4634, ×62.
    12, Male left valve, paratype Io.4633, ×62.

Figs. 8, 9. *Costa elongata* sp. nov. Campanian.   p. 54

    8, Left side, complete carapace, holotype Io.4628, ×56.
    9, Right valve, paratype Io.4629, ×60.

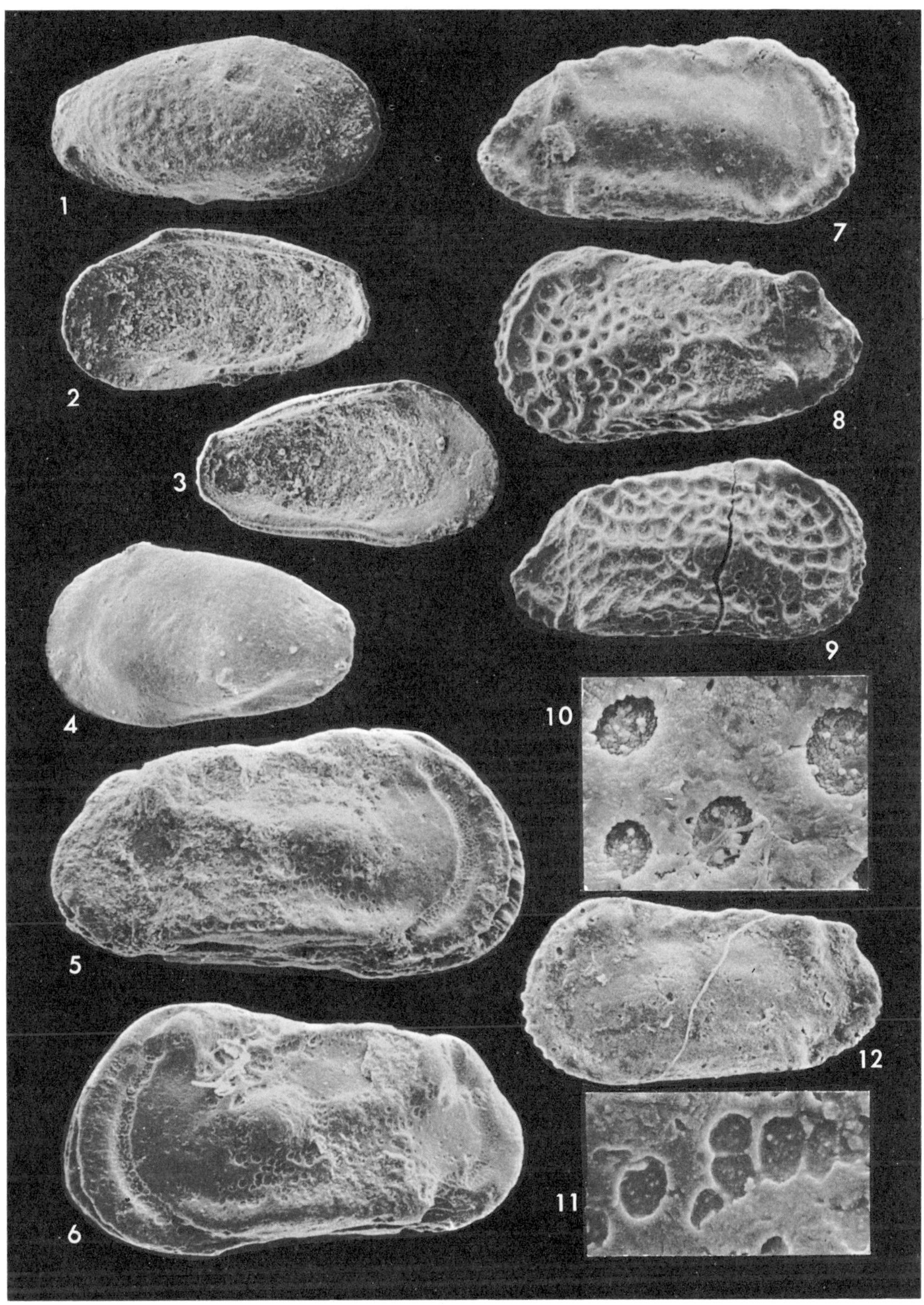

BATE, Upper Cretaceous Ostracoda, Carnarvon Basin

EXPLANATION OF PLATE 22

Figs. 1–3. *Trachyleberis anteplana* sp. nov. Campanian. p. 74

1, Central muscle scars: 4 oval adductor scars with antero-median V-shaped frontal scar; paratype Io.4617, ×620.

2, 3, Posterior and anterior hinge teeth, right valve; paratype Io.4617, ×310.

Fig. 4. *Trachyleberis* sp., Type B. Coniacian. p. 74

Right valve, Io.4613, × 100.

Fig. 5. *Trachyleberis* sp., Type A. Coniacian. p. 74

Posterior half of left valve, Io.4611, × 100.

Fig. 6. *Trachyleberidea* sp. Coniacian. p. 73

Left valve with slight damage to postero-dorsal slope; Io.4627, × 100.

Figs. 7, 8. Genus B sp. Santonian. p. 78

7, Right valve, Io.4623, ×100.

8, Right valve hinge, Io.4626, ×240.

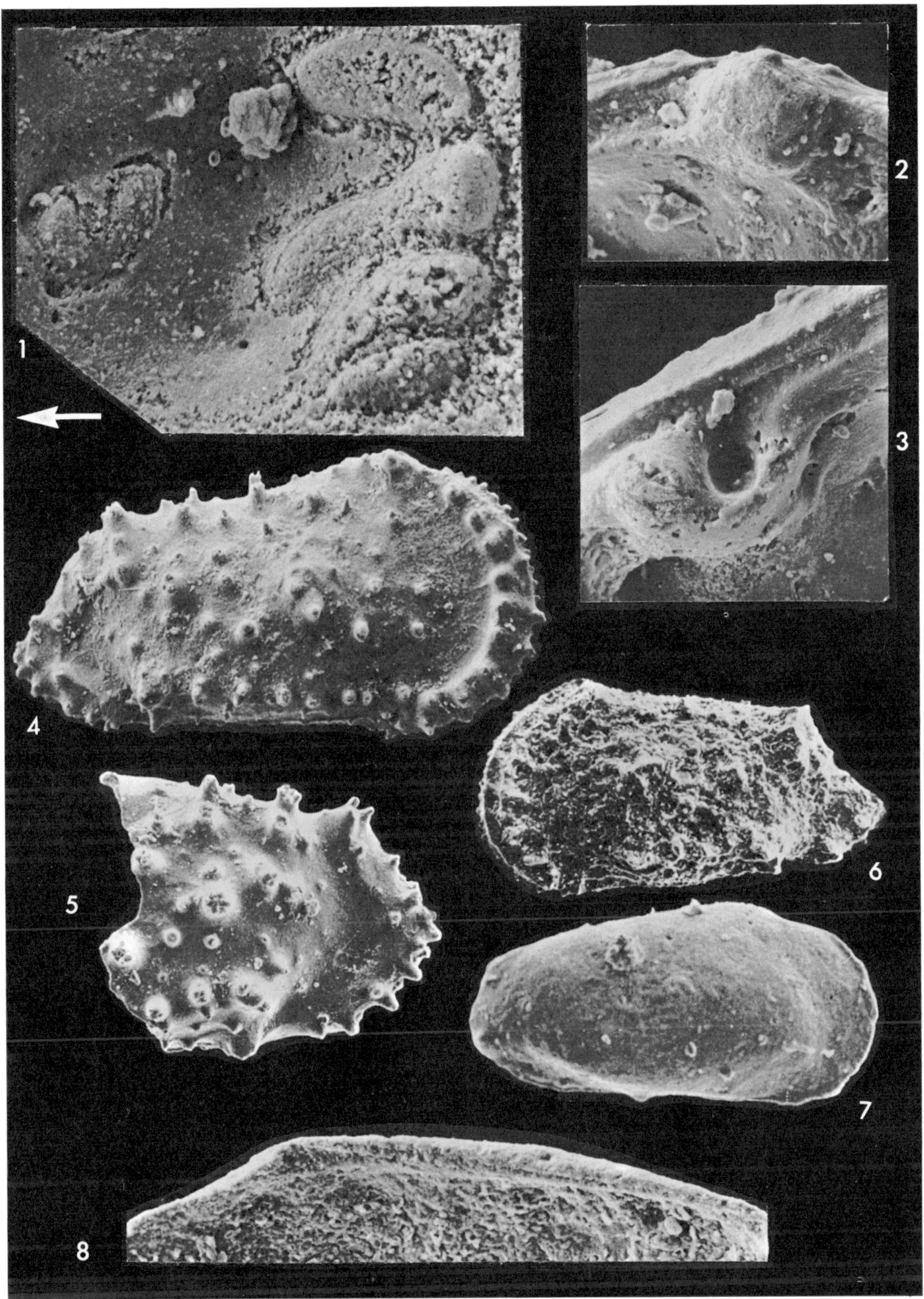

BATE, Upper Cretaceous Ostracoda, Carnarvon Basin

EXPLANATION OF PLATE 23

Figs. 1–8. *Limburgina formosa* sp. nov. Campanian. p. 64

1, Male right valve, paratype Io.4650, ×67.

2, Male left valve, holotype Io.4643, ×67.

3–6, Internal view and hinge enlargements of female left valve; note stepped posterior hinge socket, dorsal muscle scars beneath median and anterior hinge elements and opening to eye tubercle just anterior to anterior hinge socket; paratype Io.4644; fig. 3, ×67; figs. 4–6, ×620.

7, 8, Enlargement of ornament to show small ingrowing spines, small normal pore canal apertures, and conical spine bases; eye tubercle carries small spines on its flanks; holotype Io.4643; fig. 7, ×290; fig. 8, ×360.

Fig. 9. *Trachyleberis* sp., Type B. Coniacian. p. 74

Broken left valve, Io.4612, × 67.

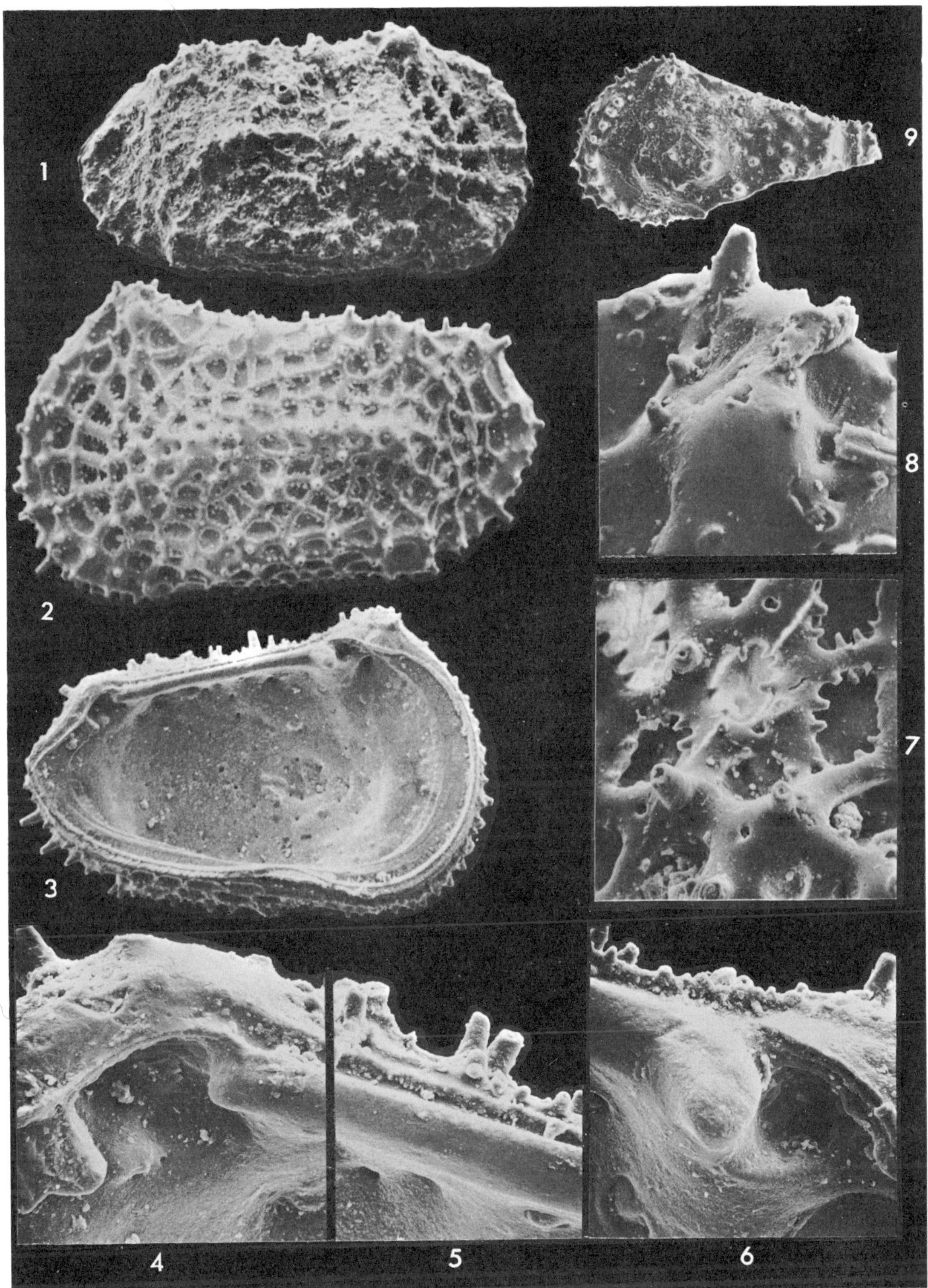

BATE, Upper Cretaceous Ostracoda, Carnarvon Basin

EXPLANATION OF PLATE 24

Fig. 1. *Oertliella exquisita* sp. nov. Campanian. p. 66

Internal view, male left valve, paratype Io.4653, × 100.

Figs. 2–6. *Hermanites sagitta* sp. nov. Campanian. p. 59

2, Female left valve, paratype Io.4640, × 100.

3, Internal view, female left valve, paratype Io.4638, × 100.

4, Internal view, male right valve, paratype Io.4641, × 100.

5, Dorsal view, female carapace, paratype Io.4636, × 100.

6, Dorsal view, female right valve, showing antero-median hinge tooth and eye tubercle; paratype Io.4639, × 240.

Fig. 7. *Hystricocythere imitata* gen. et sp. nov. Campanian. p. 76

Ornamentation of small tubercles situated on reticulation; left valve, paratype Io.4676, × 680.

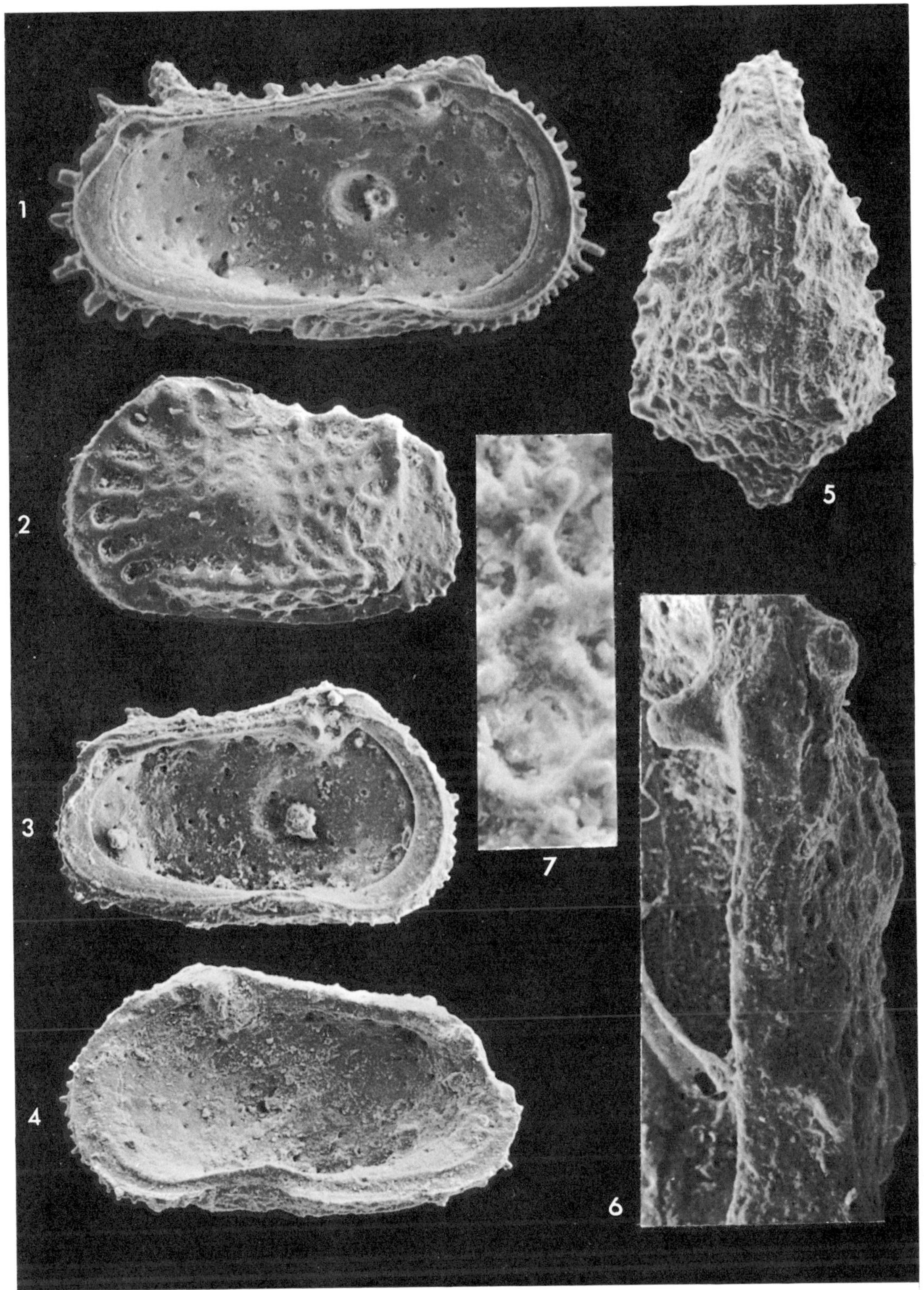

BATE, Upper Cretaceous Ostracoda, Carnarvon Basin

EXPLANATION OF PLATE 25

Figs. 1–3, 6, 7. *Oertliella exquisita* sp. nov. Campanian. p. 66

1, Female left valve, holotype Io.4651, ×100.

2, Female right valve, paratype Io.4652, ×100.

3, Internal view, male right valve, paratype Io.4654, ×100.

6, Enlargement of ornamentation to show normal pore canals opening within the pits produced by the reticulation; very small pores surround the pore canal apertures which are situated within a slightly upraised area; holotype Io.4651, ×1,100.

7, Anterior tooth, male right valve, paratype Io.4654, ×312.

Fig. 4, 5. *Hermanites sagitta* sp. nov. Campanian. p. 59

4, Female right valve, holotype Io.4635, ×100.

5, Internal view, female right valve, paratype Io.4639, ×100.

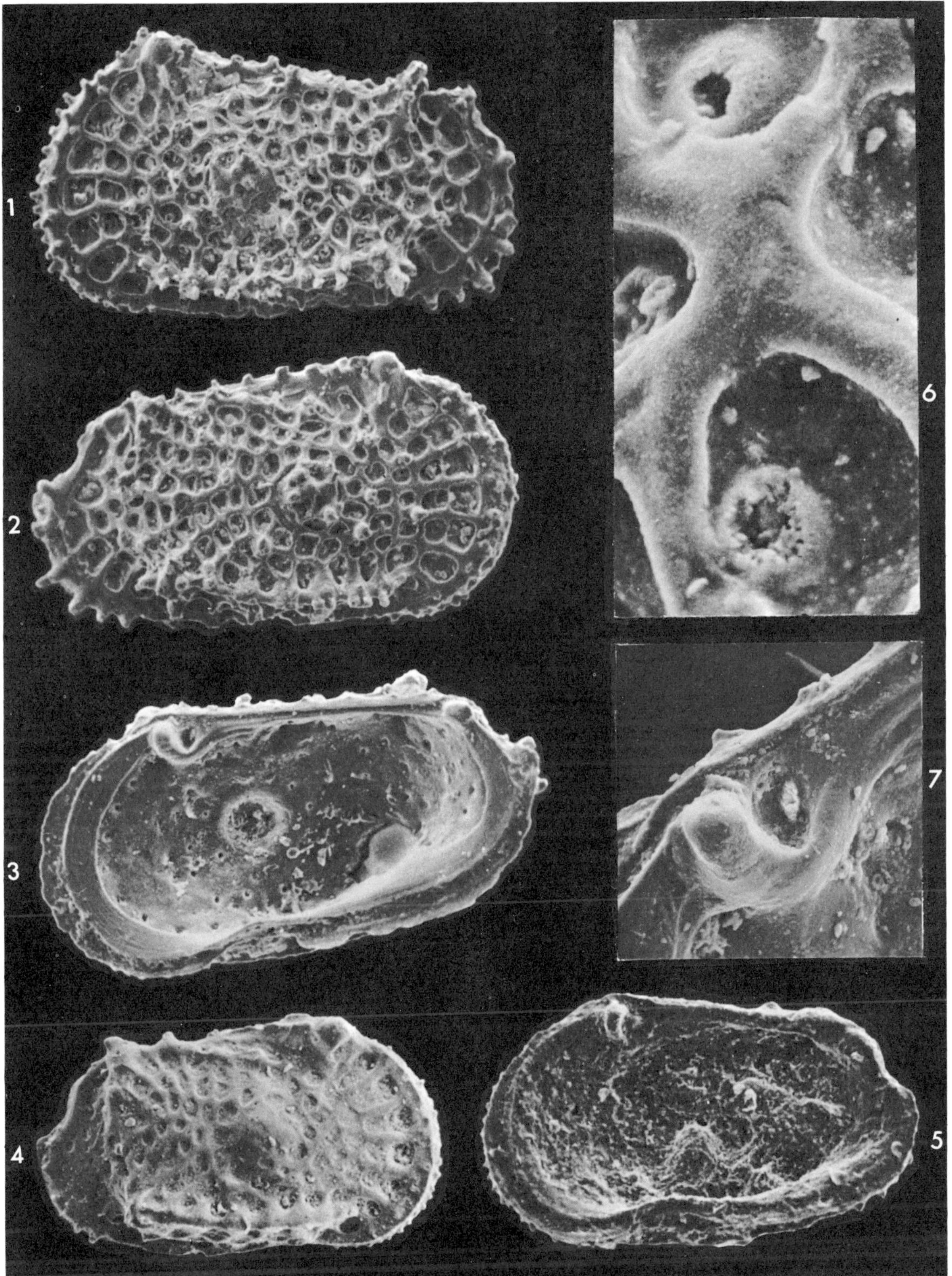

BATE, Upper Cretaceous Ostracoda, Carnarvon Basin

EXPLANATION OF PLATE 26

Figs. 1–8. *Scepticocythereis ornata* gen. et sp. nov. Santonian. p. 68

1, Male left valve, holotype Io.4658, × 100.
2, Internal view, male left valve, paratype Io.4667, × 100.
3, Dorsal view, male carapace, paratype Io.4659, × 100.
4, Juvenile right valve, paratype Io.4668, × 100.
5, Juvenile left valve, paratype Io.4663, × 100.
6–8, Internal view of juvenile right valve with enlargements of posterior and anterior hinge teeth; paratype Io.4662; fig. 6, × 100; figs. 7, 8, × 295.

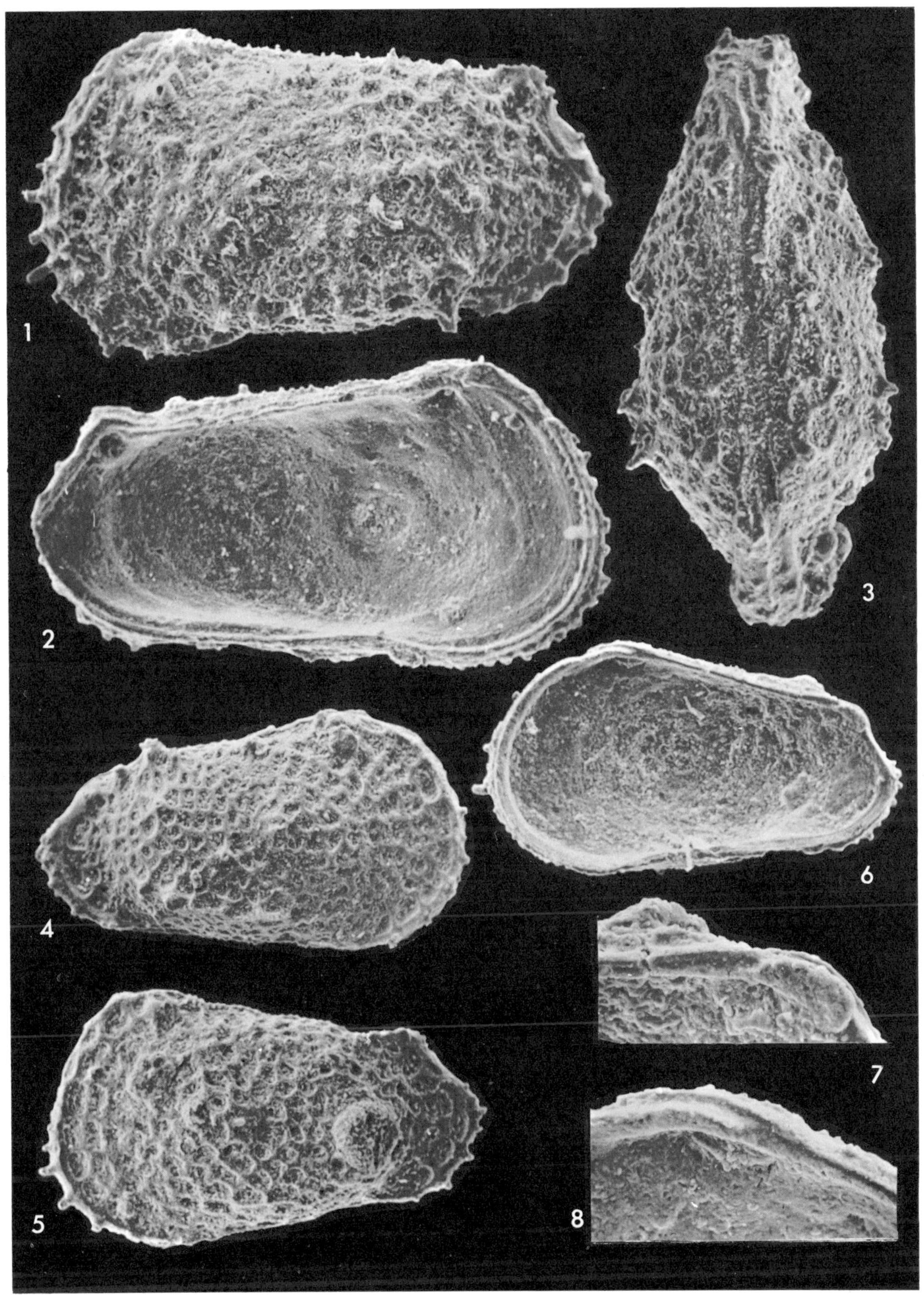

BATE, Upper Cretaceous Ostracoda, Carnarvon Basin

EXPLANATION OF PLATE 27

Figs. 1–10. *Hystricocythere imitata* gen. et sp. nov. Santonian, Campanian. p. 76

1, 7, External and internal views, right valve, holotype Io. 4669, Santonian, × 100.

2, Right valve almost identical to holotype but with slight convexity at postero-dorsal slope; paratype Io.4674, Santonian, × 100.

3, 8, Dorsal and ventral views, complete carapace, paratype Io.4672, Santonian, × 100.

4, 10, External and internal views, left valve, paratype Io.4670, Santonian, × 90.

5, Internal view, left valve, paratype Io.4671, Santonian, × 100.

6, Left valve of slightly elongate form; paratype Io.4675, Santonian, × 100.

9, Left valve of quadrate form, paratype Io.4676, Campanian, × 100.

Figs. 11, 12. *Scepticocythereis ornata* gen. et sp. nov. Santonian. p. 68

11, Female right valve, paratype Io.4664, × 100.

12, Ventral view, female carapace, paratype Io.4665, × 100.

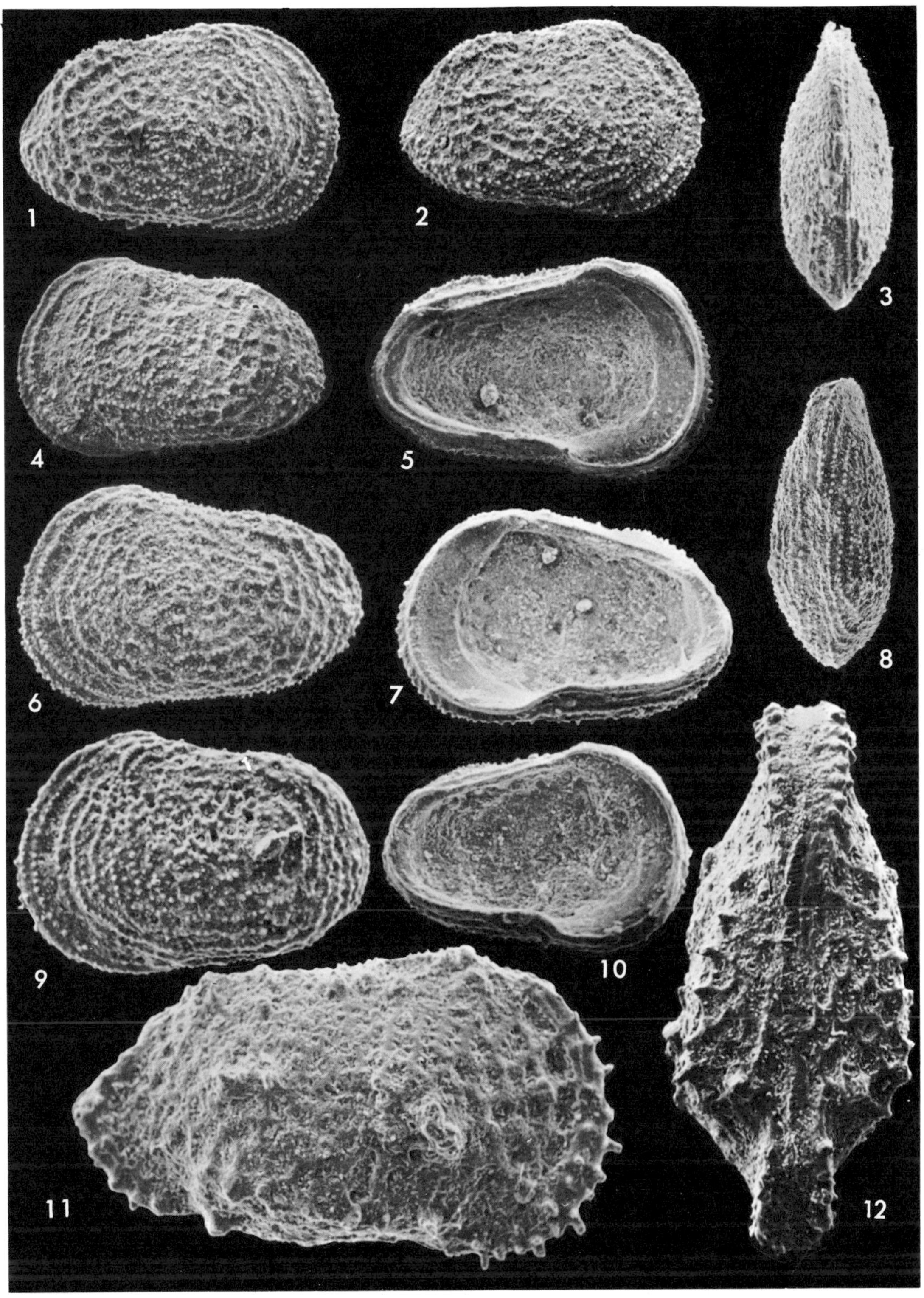

BATE, Upper Cretaceous Ostracoda, Carnarvon Basin